THE MYTH GAP

www.penguin.co.uk

THE
MYTH

WHAT HAPPENS WHEN EVIDENCE
AND ARGUMENTS AREN'T ENOUGH?

GAP

FOREWORD BY TIM SMIT

ALEX EVANS

eden project books

TRANSWORLD PUBLISHERS
61–63 Uxbridge Road, London W5 5SA
www.penguin.co.uk

Transworld is part of the Penguin Random House group of companies
whose addresses can be found at global.penguinrandomhouse.com

Penguin
Random House
UK

First published in Great Britain in 2017 by Eden Project Books
an imprint of Transworld Publishers

A CIP catalogue record for this book
is available from the British Library.

ISBN 9781909513112

Typeset in 11.25/14.25 pt Minion Pro by Jouve (UK), Milton Keynes
Printed and bound in Great Britain by Clays Ltd, Bungay, Suffolk

Penguin Random House is committed to a sustainable
future for our business, our readers and our planet. This book
is made from Forest Stewardship Council® certified paper.

MIX
Paper from
responsible sources
FSC® C018179

5 7 9 10 8 6 4

For Isabel and Kit

CONTENTS

CONTENTS

FOREWORD

Imagine if you will that you find yourself crossing one of Britain's great Victorian railway stations – St Pancras, maybe – when suddenly the roof transforms into a magnificent single-span domed plaster ceiling. We passengers are frozen mid-movement, while above our heads a giant fresco starts to appear, bearing the faces of all our descendants from the next one hundred years. Suddenly the faces start to speak to us. What do they say? What, through their trans-generational wormhole, do they wish us to do? To act upon?

It is a strange thought, but I often wonder whether the intellectual giants who spawned the Renaissance or the Age of Enlightenment were aware that they were at the start of a defining period. Anyone who reads Humphrey Jennings' magisterial *Pandaemonium 1660–1886: The Coming of the*

Machine as Seen by Contemporary Observers – the book that inspired Danny Boyle's London 2012 Olympic opening ceremony – will be struck by the ever-growing fever for knowledge as the commentators' correspondence is transcribed, their excitement pouring from every page. But there is also a kind of madness for control and efficiency, and behind it the fragmentation of both society and thought into siloed, easily manageable pieces. Nature under the thumb.

Elsewhere, the poetry of the late eighteenth century is shaved of its whole-world subject matter, a pale shadow of the grand Homeric model in which matters of religion, kingship and morality were meat and drink, alongside the dramas and passions of everyday life. Instead of those grand myths which had given shape to the Classical world, poets' territory becomes steadily more constrained to the personal and subjective.

A trivial point, one might say, but once the influence that emotions and society can bring to bear on behaviour and community cohesion is dismissed, we inexorably slide towards where we are today. Now, we live in an age of unaccountable capital where measures of success are shorn of moral compass, where the constructs of man have been accorded the status of natural laws. False idols indeed.

But we are also living at the start of a new Age of Enlightenment. One in which there has been more invention over the last seventeen years than in all history before that. We stand on the edge of discoveries that, for good or ill, will change how we live on this one planet of ours, when we will finally discover whether the name we gave ourselves – *Homo sapiens sapiens* (the wise wise hominid) – is ironic or apt. We are creatures of story, at our strongest when we tell of how the forces of good can conquer evil, and at our weakest when we allow others to tell our story for us in terms of false virtues and little more than what makes us content.

Alex Evans has written a marvellous exploration of how the roadmap of story could be read, and an unsparing critique of how, in the corridors of power, it is possible to become removed from the warmth of the hearth and influenced instead by the heat of the furnace. *The Myth Gap* points to a truth we all must learn: that something big, something that could change the world into a much better place, can emerge from the marriage of millions of small things, connecting, weaving in and out with a natural pattern, giving meaning to the whole.

So as we break the spell on the station concourse and listen to the chorus of our descendants, we can reflect that the myth by which we have lived until now – that the purpose of evolution was to end up with us – may well have progressed. I suspect that our chorus may use words relating to nature and stewardship. And I hope it will point us towards the corner for the poets, wordsmiths and storytellers: for undoubtedly, on this new roadmap of ours, it is that way that redemption lies.

Tim Smit

INTRODUCTION

THIS IS A BOOK ABOUT STORIES. Stories about people's relationships with each other and with the world around them; stories about periods of crisis and transition in history; stories about what happens when things get broken at the deepest level, and how to mend them. The kind of stories, in other words, that we call myths.

It's also my own story. Almost for as long as I can remember, I wanted to work on global issues. And once I'd seen my first episode of *The West Wing*, I knew I wanted to do it as a political adviser inside government.

I got my wish. In 2003, I became one of two special advisers to the British Secretary of State for International Development (first Valerie Amos, and then Hilary Benn) in Tony Blair's government. I did three years in the job, served as part of Labour's general election 'war room' in 2005, and sat in on meetings with everyone from Bob Geldof to then United Nations Secretary-General Kofi Annan.

My abiding impression was one that took me by surprise: how little power the government really had, at least on the big global issues. Take climate change, the issue I knew best and had worked on for longest. For all his faults, Tony Blair was a leader who really did care about climate change – to the extent of making it one of two centrepiece issues at the G8 summit that the UK hosted in 2005, which I worked on. Yet for all that, he barely influenced the global agenda at all.

At the time, I figured that the main problem was George W. Bush's administration in the United States – and that once Bush was out of office, there would be a huge window of opportunity to get serious not just on climate change but on all of today's defining global issues. So in 2006, I left government and went to work at the Center on International Cooperation, a small foreign-policy think tank at New York University known for working closely with the UN.

Almost as soon as I'd started, I found myself seconded to the UN Secretary-General's office to help organize the UN's first head-of-government-level climate summit. This was straight after Hurricane Katrina and the global warming documentary *An Inconvenient Truth*. You could *feel* the mood of purpose and resolve in the Secretary-General's office on the thirty-eighth floor of the UN headquarters, and the sense of momentum towards the 2009 Copenhagen climate summit, set for two years later.

When the financial crisis, together with a dramatic global oil and food price spike, hit a year or so later in 2008 – just as Barack Obama became the Democratic candidate in the US presidential election – it seemed as if we were at a moment full of possibility for real change.

True, the abject failure of the Copenhagen climate summit in 2009 dented my enthusiasm a bit. But looking at it another way, I saw its failure as a vindication of something that I'd been arguing for a long time: that if we wanted to get better at managing an interdependent world, then we needed to face up to the massive issues of fairness and inequality bound up with our economy's unsustainability, rather than forever sweeping them under the carpet.

I wasn't just thinking about the fact that it's always poor people who are the first casualties of climate change or environmental degradation, or who become refugees when resource scarcity contributes to increasing risks of violent conflict or extreme weather events.

Even more fundamentally, I had in mind the question of who gets to consume what at the point when the global economy starts hitting its environmental limits. After all, the single biggest challenge of the twenty-first century is the speed at which human activity is approaching a whole host of planetary boundaries. Not just limits to the atmosphere's ability to absorb our emissions, but limits also to fresh-water use, availability of land, how much biodiversity loss the Earth can cope with, and how much our oceans can acidify before they reach irreversible tipping points.

In such circumstances, questions of fair shares become unavoidable – because as Gandhi put it, 'There is enough for everyone's need, but not for everyone's greed.' And climate change, of course, is the example par excellence, where solving the problem means figuring out how to share a finite global-emissions budget between 7 billion people.

But the same challenge also applies to unsustainable consumption more broadly. If the average Briton or American uses far more than their fair share of the atmosphere's capacity to absorb carbon emissions, then the same goes for their use of energy, arable and grazing land, fresh water, fisheries, timber, fertilizer, and so on. In the interlocking series of crises we now face as a result of overconsumption across the board, some of us bear much greater responsibility than others.

And, of course, *within* the UK or the US, too, some citizens consume much more than others – just as they do in every other country around the world, as inequality between the world's rich and poor grows. Increasingly, our economy is following a 'winner takes all' pattern in which a few people excel and receive disproportionate rewards, while far more people grapple with economic insecurity, low pay, or unemployment.[1]

It's not hard to make out ways in which these trends are

contributing to marked discontent and mistrust in political systems – now more than ever, in the wake of the UK's Brexit referendum and Donald Trump's stunning victory in the race for the US presidency. What with a billion people still mired in absolute poverty and increasingly seeing migration as their way to a better life, and a future of climate change and resource scarcity in prospect, I've long felt that what our policymakers most need to do is to step back and look at the bigger picture, recognize the crisis of unsustainability rapidly approaching, and then take resolute action to lead their societies in a momentous change of course.

So it felt like everything I'd been waiting for when, amid the ashes of the Copenhagen summit in early 2010, UN Secretary-General Ban Ki-moon announced the creation of what would become the UN High Level Panel on Global Sustainability, to come up with a new way forward on issues such as poverty, climate, inequality and environmental protection. I was asked to come on board as the panel's writer, the person charged with finding the right words to express the collective view of its members.

At the time, I felt giddy with excitement. I'd been pushing the Secretary-General's office to set up a body like this for the last five years. I felt sure that if we could just pull together a group of senior policymakers and remove them from the day-to-day grind of crisis management, they'd have a chance to get right to the nub of the problem –

and then we'd have a core of leaders willing to champion whatever it would take to solve these issues.

So I thought, right up until the moment when it all fell apart. It was 2011. I was in a small room at the UN's headquarters in New York. Seated around the table were the panel's members: prime ministers, presidents, foreign ministers and ambassadors, from the United States, the EU, China, Brazil, India, and a dozen other countries.

It was the panel's final meeting. This was where it was all supposed to come together. This was the moment when the panel's members would rise above the endless bickering of G8s, G20s, G77s and all the other international groups of 'like-minded' countries, and start to think and act like a 'G1': a group, in other words, that would recognize that we all live on one planet and need to act accordingly.

What I actually saw that day was instead an illustration of the 'G Zero': a world in which *no* leaders are prepared to think beyond their national interests or show vision on tough global issues.[2] Just like at Copenhagen two years earlier, the US joined forces with emerging economies like Brazil to block anything remotely ambitious.

As the last vestiges of my naivety fell away, it dawned on me that, far from engaging with the fundamental issues of fairness that arise in the context of environmental limits, the High Level Panel would in fact barely be willing to acknowledge the existence of any natural limits in the first place.

And, amid my disappointment and disillusionment, I lost the faith that had sustained me through a decade and a half of work as a policy geek: the conviction that rational arguments, backed up by well-presented evidence, would be enough to persuade politicians of the radical actions needed to build a fairer and more sustainable world.

Which made me wonder: if evidence and rational arguments aren't enough, then what is?

I've thought about that a lot since the High Level Panel finished its work in 2011, and especially after I moved with my family to Ethiopia the following year for my wife Emma's job in international development, which put some helpful distance between me and the policymaking worlds of London and New York. It was a shock suddenly to see the issues I'd been writing about for a decade or more right in front of me – refugees, droughts, food crises – and to have to think hard about how unsustainable lifestyles like mine are exacerbating these challenges.

It was also a strange and unsettling sensation to watch from afar as Donald Trump became the Republican presidential nominee and as the Brexit campaign triumphed with Britain voting to leave the European Union. In both cases, fears about economic security and migration were played on, while hugely resonant stories were powerfully set out about building walls and taking back control from remote elites.

Today, I feel more sure than ever that if we're to

overcome issues on the scale of the ones that confront us today, we need to look far beyond policymakers and pie charts.

Instead, I've come to believe that we need to build the kind of mass movement that has in the past created the political space to end slavery, or establish new civil rights, or secure the write-off of billions of dollars of Third World debt. Today, we need a movement that will push for what my colleague Rich Gower and I have called a 'restorative economy': one that lives within environmental limits, ensures everyone can meet their basic needs, and keeps inequality within reasonable limits.[3]

To animate a movement on this scale, we need powerfully resonant stories, and they need to be stories that unite rather than divide us. Not so long ago, our society was rich in these kinds of stories, and we called them myths. Today, though, we have a myth gap. And to fill it, we need new myths that speak about who we are and the world we inhabit and help us work through our grief for what is happening all around us, both to the natural world and to people. And they need to give us hope for the future, by moving beyond the arid jargon of 'sustainable development' and instead telling us stories of *restoration*: how we can repair the damage we've done to the climate, help to mend the ecosystems we've broken, and right the wrongs being done to other human beings by how we've organized our economy.

In a world of 7 billion people, there's clearly no single myth or set of myths that will work for all of us. Instead, we *all* need to discover and develop our capacity to tell stories: not just on our own, but together; not just as a way to pass the time, but as the motivator for a momentous change in how seven billion of us live together on Earth.

PART ONE

THE FRONT LINE

1

How the climate movement rebooted itself

IT'S THE AFTERNOON of 23 June 1988. NASA scientist Jim Hansen is testifying on global warming to the US Congress. Outside, it's a sauna. Temperatures are sweltering to an unheard-of high of 38 degrees Celsius. The legislators and journalists in the room are close to fainting.

It's one of those moments when everything comes together. The next day, climate change headlines the *New York Times*.[1] By September, 58 per cent of Americans have heard of it.[2] Two months after that, the Intergovernmental Panel on Climate Change (IPCC) is set up. The global fight against climate change has just begun.

Now it's 1990. In Geneva, the Second World Climate Conference is taking place. British prime minister Margaret Thatcher is lavishing praise on the IPCC, which has just published its first Assessment Report. She's leading not with heartfelt pleas to save the Earth or passionate calls to arms, but instead with rational arguments about self-interest, telling leaders that 'it may be cheaper or more cost-effective to take action now than to wait and find we have to pay much more later'.[3]

The terms on which climate policy will play out over the next two decades have now been set. This is to be one of the most technocratic agendas the world has ever seen. Climate change will be owned by a 'priesthood' of experts, with its own language, rituals, gatherings, assumptions and, above all, abbreviations. It's a thoroughly insider game, and the ragtag bands of activists outside the summits are a charming irrelevance. As for the public, their job is to listen to the experts and then remember to turn out the lights. It's certainly not to *participate*, much less wield *power*.

It all looks very different in global development campaigning, where a retired lecturer called Martin Dent has just had the idea of linking the ancient concept of Jubilee – under which all debts were supposed to be cancelled every half-century – to a campaign to write off Third World debt. His idea – known as Jubilee 2000 – lights a spark that will, within a decade and a half, lead to the forgiveness of all $40 billion owed by the most highly indebted poor countries to the World Bank and the International Monetary Fund. Where the Jubilee 2000 debt-relief movement will be driven by mass mobilizations and moral outrage, climate policy is instead about science, summits and economic self-interest.

So it's probably no surprise, then, that when we look at opinion polls from fifteen years later, in 2005, we find the public in developed countries saying they believe climate change is real, urgent and human-caused – but also that they are not particularly willing to countenance big lifestyle changes as a result, or to demand radical action from politicians.

It's a classic 'thin yes'. There's no great conviction or urgency here. Politicians understand this, so the policies they adopt are similarly half-hearted. The US pulls out of the Kyoto Protocol, the world's first serious attempt to tackle climate change. Canada and Australia follow suit. The European Union stays in, but fills its new emissions trading scheme with loopholes. Emerging economies like China and India hang back, secure in the knowledge that there will be no pressure on them to act as long as the rich world is so clearly failing to take the lead.

The UN's annual cycle of climate talks looks increasingly like a multilateral zombie: staggering slowly forward, never really getting anywhere, never quite dying either. And all the while, the world's emissions keep rising and rising and rising – by more than 60 per cent between 1992 and 2012.

And then suddenly everything changes.

In 2009, environmental NGOs are still running with

their same old theory of change. It seems to be going well. A US presidential campaign has just been fought between two candidates, John McCain and Barack Obama, who both agree that climate change is real, human-caused and urgent. Better still, the House of Representatives has passed a climate bill – the US's first. Now it only needs to clear the Senate. That should be a walk in the park, perfectly teeing up December 2009's UN climate summit in Copenhagen.[4]

Few have yet noticed a new phenomenon called the Tea Party, which proceeds to spend the summer 'invading town halls, dominating talk radio and Fox News, and generally scaring the bejesus out of Republican legislators', as climate journalist David Roberts puts it.[5] And it turns out that destroying climate legislation is at the very top of the Tea Party's hit list.

By the time the Senate comes back from its summer break, prospects for US climate legislation have fallen apart. Six months later, far from agreeing a deal for a maximum 2-degree-Celsius global temperature rise, as the NGOs had hoped and expected, the Copenhagen summit disintegrates, having got rid of legally binding targets altogether. Instead, from now on global climate policy will be voluntary. Once the pledges made at the summit are totted up, it emerges that the world is now on course for a rise of between 3.6 and 5.3 degrees: well into the disaster zone.[6]

As environmental NGOs retreat to lick their wounds, they commission Harvard political scientist Theda Skocpol to write a report explaining what the hell just happened. Her conclusion is that the Tea Party played an outsider, populist, values-led game against the NGOs' insider, technocratic, facts-led game – and ran over them with a tank. The NGOs never saw what was coming until it was far too late. Their polling results had looked great, right up until the moment when it all imploded. As David Roberts notes:

> National polls tell enviros what they want to hear: in the abstract, majorities always support clean air and clean energy. Enviros mistook these results for constituencies. But poll results do not attend town halls or write members of Congress or exhort their fellow citizens through ideological media. Constituencies do that.[7]

If the story of how climate activists were defeated by the Tea Party in 2009 seems familiar, that's because 2016 offered two re-runs of the same underlying dynamic. One was Britain's referendum on its membership of the European Union. The Remain campaign ran with exactly the same kind of technocratic, fact-heavy tactics that climate campaigners had used in 2009. The Vote Leave campaign, on the other hand, played fast and loose with the facts but set out by far the more powerful story, presenting Brexit as an insurgent movement fighting remote, unaccountable elites.

On the other side of the Atlantic, meanwhile, Donald

Trump successfully employed exactly the same playbook against his challengers for the Republican nomination – before going on to use the same strategy to beat Hillary Clinton, who adopted a far more dry and technocratic style of campaigning, in the election itself.

But now jump to summer 2014. Climate activists have just mobilized nearly half a million people on to the streets of New York. Over the next year and a half, the media is saturated with totemic images: people taking to the seas in kayaks in America's Pacific northwest to stop oil rigs from being towed out to the Arctic, or surging relentlessly through lines of riot police to shut down production at a German coal mine.

President Obama seems to be making a new climate speech every month. The fossil-fuel-divestment campaign – which aims to push big investors to boycott fossil-fuel companies, much as the anti-apartheid movement pressured companies to pull out of South Africa – claims one scalp after another. First Oxford University's endowment fund, then Axa insurance, the Church of England, even Norway's vast sovereign wealth fund (created, ironically, to invest the proceeds of its oil production). The Bank of England starts an inquiry into the financial risk of oil companies becoming 'stranded assets' as their reserves become officially un-burnable.

To cap it all, in December 2015, the COP21 climate summit in Paris ends with climate activists in a state of stunned euphoria after governments agree to a new target of zero global emissions by the early second half of the century. Instead of hanging back, the emerging economies whose emissions are growing fastest have come forward with ambitious new action plans. Most surprisingly of all, governments have set a new target of limiting global warming to 1.5 degrees Celsius – a far more ambitious goal than their previous target of 2 degrees.

Something massive shifted in the politics of climate change between 2009 and 2015. But what, exactly? And what can that shift tell us about how to unlock far-reaching change on global issues more broadly?

2

What climate activists learned from 2009

WHEN I MEET Bill McKibben, the godfather of climate activism, at the Paris talks in 2015, the successful outcome is not yet a done deal. He's holed up in the summit's media centre with his crew from 350.org, the shock troops of the new global climate movement, together with their allies from Avaaz, a global campaigning movement with more than 40 million members.

Mobile phones buzz incessantly as activists plan a big action for the summit's final weekend. Small huddles of campaigners and policy experts are discussing tactics for how to push countries to join a 'high ambition coalition' that they're assembling. At the centre of it all is McKibben: laconic, watchful, fierce.

For all McKibben's bookishness – he's a writer by trade,

and peppers his arguments with numbers – there's also another side to him: a quiet fury that's like nothing so much as an Old Testament prophet railing against iniquity. ('These are rogue companies. They're out of control. They have way more gas and oil than we can safely burn. They're *immoral*.')

This kind of talk is a long way from the evidence-based, PowerPoint-presented arguments that environmental NGOs were making until 2009, and McKibben knows it. 'What doesn't work for movement building is the endless repetition of paragraphs and pie charts and all the things that people like to write in reports,' he says. McKibben epitomizes the shift that took place in climate campaigning after 2009 – when climate activists learned three crucial lessons from their defeat.

The first thing they realized was that they had to *build a mass movement*. As McKibben puts it to me:

> There really wasn't a mass movement around climate before Copenhagen. I mean, we had our first day of action six weeks before Copenhagen. The point was we were starting to fill this vacuum. And yeah – the inside game clearly hadn't worked. They didn't have the muscle to make it happen. Copenhagen was a complete failure. So we had to get power some place.

He's not the only commentator to observe that there's now a climate movement where there wasn't one before.

Veteran activist Todd Gitlin has written that what makes a movement is simply enough people feeling a part of it – sensing a shared culture, and forcing those watching to take note and take sides.[1] On that basis, he observes, 'There is today a climate movement as there was a civil rights movement and an anti-war movement and a women's liberation movement and a gay rights movement.'

The birth of a genuinely global climate movement is a big deal, and not just because of its power to demand that policymakers do stuff. More subtly, history shows that movements have a special ability to propagate new values.

Much as we like to think of ourselves as rational beings, pollsters and psychologists know better. When it comes to how we make up our minds about political issues, it turns out that evidence, facts and data matter much less than the values held by the people we hang out with: family, friends, colleagues. (For proof, look no further than your Facebook wall: ever noticed how many of your friends think broadly the same way you do about political issues?)

And this is where movements can be so powerful and disruptive, not just in telling new stories, but in creating 'congregational spaces' where these stories can be incubated and lived out.[2] Movements have, as Princeton professor Thomas Rochon puts it, 'a uniquely powerful ability . . . to create controversy about ideas that were once consensus values'.[3] They're the perfect environment for stirring things up and challenging old orthodoxies. And suddenly, climate activists are doing exactly that.

The second thing climate campaigners learned after 2009 was to *build their movement around small groups* and on an open-source model – a big contrast with the kind of centralized, Bono- and Coldplay-led campaigns that global development campaigners have largely focused on in recent years. As McKibben observes, '350 set out to be a very explicitly open-source movement: here's the logo, do with it what you will.'

This, too, recognized and embedded a crucial lesson about how to make change happen. Mass mobilizations around pop concerts are great for making noise at a G7 or UN summit – but if you're playing a long game, as the climate activists now are, and you want to keep the pressure up over time, then small groups are the glue that holds everything together.

Military strategists have always understood this, which is why armies are organized around the basic unit of platoons. The Jubilee 2000 debt-relief movement understood it too, and built its campaign largely from church congregations. The Obama '08 campaign won using a ground operation based on small groups. And now climate activists are doing it again – whether on fossil fuel divestment, a movement rooted in college groups across the US, or in the small-group tactics that underpin mass actions like shutting down production at the Ende Gelände coal mine, Germany's largest.[4]

Third and most importantly, *climate activists now have a terrific story*. If values trump facts and movements shift values, then a really resonant story is the spark that lights the movement's flame.

Today, the climate movement is following in the footsteps of anti-slavery abolitionists two centuries ago, civil rights advocates fifty years ago, and Third World debt campaigners a decade ago – all of whom framed their stories not in terms of rational self-interest but as *moral* challenges. As McKibben puts it, 'People understand the world through stories. So we end up telling an endless series of stories . . . [like] defuse the carbon bomb or it's game over for the planet.'

Nor is it just climate activists who've started framing the issue as a moral one. Pope Francis devoted an entire papal encyclical to climate change in 2015, and turned up at a high-profile UN summit on sustainable development a few months later to underline the point.[5]

You can see the shift over the course of Obama's eight years in office, too. In the first term, his language on climate was all about how many 'green jobs' would be created. But by his second term, when Obama was enacting major policy reforms on climate, it was just 'the right thing to do'. (In psychological language, this is a shift from extrinsic to intrinsic motivation, which research shows over and over again is a far more powerful basis for action.[6])

The climate movement's morally based stories tend to

have a very particular target, too: the 'rogue companies' in the oil and gas sector. When I ask McKibben what's next for 350.org after its breathtaking success in persuading Obama to halt the proposed Keystone XL pipeline from Canada to Texas, a campaign many thought was unwinnable, he replies simply: 'I want to spend a lot of time focused on Exxon.'

You can see why. When we meet, Exxon has just been subpoenaed by New York's attorney general to investigate allegations that it has suppressed climate change research and lied to the public or investors about the risks.[7] You could hardly invent a better villain – a climate version of Big Tobacco. As McKibben continues:

> I think that's a remarkable story . . . twenty-five years ago, they alone probably could have short-circuited this entire bogus quarter-century debate we've been having, simply by saying what they knew, and we'd be a long way towards fixing the problem. And instead, they just straight up lied . . . And they continue to spend more money looking for new hydrocarbons than anyone else on Earth. They're just a huge, ongoing threat.

But there's a problem with 'enemy narratives' like these. For all their power to shock, outrage and mobilize, they're taking us away from the place we need to get to.

3

The problem with enemy narratives

MEET GEORGE MARSHALL. Although he's been campaigning on climate change for as long as Bill McKibben, he couldn't be more different. Where McKibben revels in being at the centre of things, Marshall – owlish, wry, based in a tiny village in rural Wales – likes to hang out at the edges, watching where everyone's coming from. Where McKibben is passionate and full of conviction, Marshall is thoughtful and questioning. If McKibben is the general leading the charge, Marshall is the pathfinder unit operating ahead of the main force, gathering intelligence behind enemy lines.

This approach leads to frequently hilarious results in Marshall's book *Don't Even Think About It: Why Our Brains Are Wired to Ignore Climate Change*, much of which is spent hanging out with people who see the world rather differently from him. In one episode, Marshall finds himself with a crew of Tea Party activists 'cruising along Texas State Highway 71 in the largest car I have ever seen: a seven-ton Ford Excursion, a car so huge that you need to lower a step before you can even climb inside'. He's about to spend the evening with them at a fundraising potluck dinner to try to get a better understanding of why they're so adamantly opposed to action on climate change. This is a quintessentially George Marshall kind of activity.

As I discover when I chat with him, Marshall's willingness to look at climate change through other people's eyes has made him increasingly uneasy about the kind of enemy narratives that 350.org and its activists use to such powerful effect. This is, above all, because of the risk of political polarization. In the US, what you think about climate change is the single most accurate indicator of your overall political orientation. More than gun control, more than abortion, more than capital punishment, climate change as an issue is owned by 'progressives' on the left – and as such is anathema to conservatives on the right.

The problem with that, Marshall observes, is that climate change is 'far too large to be overcome without a near-total commitment across society'. Unfortunately, he continues, climate activists can often seem more interested in maintaining the moral high ground than in bringing along people who think differently. Marshall notes that the climate movement often displays contempt for the right-leaning mainstream and their concerns, talking acidly about conservatives' selfishness, greed, or stupidity – in the process ignoring the diversity that exists between, say, 'a struggling rural family, an elderly Christian on a small pension, a community shopkeeper and a Wall Street banker', all of whom are combined into a single enemy.[1]

What's more, he continues, by focusing all the blame on a few mega-corporations and billionaire climate deniers, climate activists also risk letting the rest of us off the hook:

> The missing truth, deliberately avoided in enemy narratives, is that in high-carbon societies, everyone contributes to the emissions that cause the problem . . . the real battle for mass action will be won not through enemy narratives [but through] narratives based on cooperation, mutual interests, and our common humanity.

Marshall's point isn't that we're going to solve climate change through voluntary individual action – by more of us choosing to go vegetarian, say, or ride bicycles everywhere. He gets the need for systemic change just as clearly as McKibben does.

Where his argument differs from McKibben's is in his recognition that, actually, this isn't just about a few rogue companies. Much more fundamentally, it's about whether the rest of us are willing to give our policymakers the political space to get serious – *really* serious, war-footing serious – about climate change.

The same point applies more broadly, of course, to what we're willing to demand of our politicians on other issues. At some level we know that politicians are promising us the moon when they say we can have low taxes *and* a world-class health or education system or Scandinavian levels of social equality. But when it comes to election time, how many of us would really be willing to vote for politicians who told us straight and offered either higher taxes, or worse public services, or greater inequality?

Having worked for Tony Blair and Gordon Brown in the 'war room' of Labour's 2005 general election campaign, one thing I'm certain of is that, whatever else we might say about politicians, they do care what you and I think – obsessively so. And they understand clearly that the majority of their electorate are not in fact yet serious about climate change, or about paying substantially higher taxes, or indeed accepting the implications of lower taxes. Not *really*.

The transformation we need is not simply political: we have to be willing to transform ourselves, too. This isn't just about 'ethical consumerism'. It's about how we work, and the signals we send when we take on leadership roles in our organizations. It's about what we say to our friends, colleagues and families when climate change comes up – and whether they can see us practising what we preach. It's about how we behave in every realm of life, how we think, what we value most.

The problem with enemy narratives about Exxon, the 1 per cent, or tax-dodging corporations like Starbucks or Amazon, is that while they make activists feel great, they have very limited power to drive this larger shift. It's a point echoed by Micah White, one of the key creators of the Occupy Wall Street movement, who argues bluntly that 'protest is broken'. Instead, he says:

> The main thing we need to see is activists . . . starting to think about how to make people see the world in a fundamentally different way. What I am proposing is a type of activism that focuses on creating a mental shift in people. Basically an epiphany.[2]

Marshall agrees, arguing that activists need to be willing to build bridges to people at the other end of the political spectrum, and that they can only do this if they're willing to let go of enemy narratives and embrace a very different kind of story. There's no intrinsic need for narratives to have an enemy, he observes: many myths are instead about a quest, or a challenge, or overcoming some internal weakness. This was a central preoccupation of the psychoanalyst Carl Jung, he continues, who wrote at length about the challenge for each of us of confronting our own 'Shadow': the 'greedy internal child whom we don't wish to acknowledge or recognize and who compels us to project our own unacceptable attributes on to others'.[3]

The stories that have the power to create the kind of epiphanies that Micah White talks about are a very particular kind. They're the stories that we call myths. Myths are the most fundamental narratives of all. The ones that tell us where we are, how we got here, where we're trying to go, and how to get there – and, underneath it all, who we are. Myths don't just explain the world: they explain us, too.

In the pre-modern world, as historian of religion Karen Armstrong puts it, 'mythology not only helped people to make sense of their lives but also revealed regions of the human mind that would otherwise have remained inaccessible'. In other words, she continues, it was an early form of psychology: when our ancestors told each other tales of gods and heroes, labyrinths and monsters, they 'brought to light the mysterious workings of the psyche, showing people how to cope with their own interior crises'.

Our most respected leaders have also been able to communicate at a mythic level by means of the sheer grandeur of the stories they tell and what they are able to summon forth from their listeners as a result. My great-grandfather, Charles Wilson, was Winston Churchill's doctor throughout World War Two, and was in awe of Churchill's extraordinary power of storytelling. He not only watched Churchill deliver his orations but also saw him crafting them, for Churchill would often ask my great-grandfather to sit up with him late at night on overseas trips while he grappled with the process of finding words for crucial speeches like the one he gave to the US Congress in 1941.

Reading Churchill's 1940 speeches again today, their mythic aspect is impossible to miss. Churchill was un-flinching in his description of just how bad things were: 'The whole fury and might of the enemy must very soon be turned on us . . . if we fail, then the whole world, including the United States and all that we have known and cared for, will sink into the abyss of a new Dark Age.'[4] But he married this with a deeply hopeful vision of the future – in which 'all Europe may be free and the life of the world may move forward into broad, sunlit uplands' – and, above all, with a profoundly energizing view of what his compatriots were capable of:

Let us therefore brace ourselves to our duties, and so bear ourselves that, if the British Empire and its Commonwealth last for a thousand years, men will still say, 'This was their finest hour.'[5]

So when we find ourselves faced with an issue like climate change – a monster of a challenge that presents an existential threat, a threat that all of us have at some level helped to create, and that requires a level of resolve and collective action that we've shown in the past during wartime but that often eludes us today – it's not hard to see why myths might provide just what we need.

There's only one problem. We have a myth gap.

4

The myth gap

IT MIGHT SEEM STRANGE to go to an expert on consumer brands to talk about myths. Not to Jonah Sachs. A natural storyteller and the author of a bestselling book on how to use stories in marketing, *Winning the Story Wars*, Sachs has consulted for dozens of Fortune 500 companies and helped to devise some of the most successful viral marketing campaigns of the last ten years.

He also adores myths. Unusually for a guide to marketing, his book is peppered with references to the *Bhagavad Gita* and the Old Testament. For Sachs, these are the ultimate stories, powerful and full of lessons for anyone interested in the art of making people change their minds.

But Sachs also knows that his love for myths is today a

minority pursuit. In our rational, scientific age, most of us are more interested in facts than stories, leading to the emergence of what he calls a 'myth gap'. 'Ours is the first generation not to share myth,' as he puts it to me. It's an idea that he unpacks more fully in his book, in which he observes that 'Our mythic landscape has become brittle because many people have rejected the notion of thinking symbolically – and that's unique to our rationalist modern society.'

Nowadays, amid pitched battles between science and religion, our sole focus is often on whether things are literally true or not: 'Either the world is six thousand years old or several billion. Either we are descended from ape-like ancestors or created by God. To the modern mind, only one can be right, and the other is wrong.' Yet to the mythological mind, Sachs continues, both of these can easily be true. 'Held to the standard of literal truth, traditional myths start to crack,' he concludes.[1]

Religions have been coming off worse in this war of competing world views, leading to one of the stranger results of our rational age: religious believers claiming – disastrously – that their myths should be taken as literal facts rather than as symbolic truths. Today, the religious congregations growing fastest are those that take this approach, especially in developing countries where fundamentalist versions of Christianity, Islam, Hinduism and even Buddhism are thriving.

Worse still, as Karen Armstrong documents in her magisterial history of fundamentalism, *The Battle for God*, the fundamentalists have in recent decades become steadily more political: from the 'moral majority' of the religious right in the US and settler movements in Israel, to Salafist groups like ISIS or Boko Haram across the Middle East and beyond.[2] As a result, many of the people watching all this from the sidelines have come to see religion in general as a force for bigotry and intolerance at best, and blood-soaked ideological conflict at worst.

Against this backdrop, one of the big stories of the modern age (and especially since the mid-twentieth century) has been of people simply disengaging from religion. Religious observance has been declining steadily for decades in most developed countries – and it has done so precipitously among millennials, almost certainly the least religiously observant generation since humans invented agriculture.

All this has had a big impact on the place of myths in our society. As Sachs notes, myths can't do their job from the fringes of society. Where ancient societies were able to turn to shamans and sages for a combination of story, explanation and deeper meaning, these kinds of sources have now either lost their legitimacy or simply disappeared. This, he continues, has given rise to a situation that's both unprecedented and hazardous. As Carl Jung wrote just before World War One:

The man who thinks he can live without myth or outside it is an exception. He is like one uprooted, having no true link with either the past, or the ancestral life within him, or yet with contemporary society.[3]

Sachs concludes that 'societies without myth might easily fall apart. And a world full of such mythless men is a dangerous place.' If that sounds abstract, consider the scale of 'culture wars' in many developed countries, and the collapse in trust in leaders and institutions of all kinds – not just religious ones, but also those in politics, business and the media.

So what, if anything, has emerged to fill the myth gap? In a world in which trust is collapsing, who *do* we look to to explain what life is really about? Sachs's answer is as unexpected as it is refreshing in its candour. We've turned to 'a new kind of myth-maker – the marketer'. And while Sachs wants his fellow marketers to use their power to change the world for the better, he recognizes that that's not what most of them have been up to since their discipline was born:

From its most primitive days on, marketing has been about allowing a product or service to confer *meaning* on the purchaser. Even our take-out coffee container is a signifier of meaning and belonging. Looking at the cup, we know what tribe we are part of – Starbucks paper cup? Mainstream good-lifers or aspiring to it. Dunkin's Styrofoam? Downscale and proud. Reusable steel travel cup? Eco-conscious and responsible.

And ritual? Well, of course. What's the good of a marketing story if it doesn't give audiences a way to live that story out? You might say that introducing a new ritual is the basis of every marketing campaign. Shopping has become the single ritual we universally share.

So how is the myth that 'you are what you buy' working out for us? Not great in terms of climate change – you only have to look at the near-perfect correlation between countries' per capita GDP levels and their CO_2 emissions to see that.

Nor is consumerism doing the planet many favours in terms of environmental issues more broadly. Each year, campaigners now mark 'Earth overshoot day': the date when humanity's demand for ecological resources and services exceeds what the Earth can regenerate in that year.[4] In 2016, the date fell on 8 August – meaning that the world's human inhabitants had used up the resources that the planet can sustainably provide for a year in a little over seven months.[5]

But more subtly, our new myths don't appear to be doing a great job of making us happy either. At its core, as Joseph Heath and Andrew Potter note in their book *The Rebel Sell*, consumerism seems to be a 'product of consumers trying to outdo one another'. If consumers were just conformists, they observe, then everyone would go out and buy exactly the same stuff; there would be no reason to buy anything new. Conformism, then, fails to explain the compulsive nature of consumer behaviour – why people keep spending more and more, even though they are over-extended, and even though it brings them no more happiness in the long run.[6]

Instead, they argue, consumerism is really about all of us wanting to be different, to stand out from the crowd – an observation that 'anyone working in advertising will find crushingly obvious'. It's about status. And the problem with that, they continue, is that:

> Status is an intrinsically zero sum game. In order for one person to win, someone else must lose. Moving up necessarily involves bumping someone else – or everyone else – down ... Thus, as society as a whole grows wealthier, consumer behaviour increasingly acquires the structure of an arms race. It's like turning up your stereo in order to drown out the neighbour's music.

Even people who just want to maintain a 'respectable' standard of living wind up having to spend more in a form of 'defensive consumption'. Over recent decades, social scientists have found that people's perceptions of the absolute minimum required to live a decent life have been rising steadily, tracking the overall rate of economic growth – so that 'even the very poor are chasing a moving target'.

As Annie Leonard, the creator of the film *The Story of Stuff*, summarizes: 'We're destroying the planet, and not even having fun doing it.'[7]

5

Collapsitarianism

BUT IF MARKETERS have created a new myth to fill the contemporary myth gap, they're not the only ones.

It was late in 2011, and I was in Brussels to speak at a conference on resource security. At the time, the issue couldn't have been more topical. Global food prices had spiked to an unprecedented peak – the second time they'd done so in just three years. News bulletins were full of stories about huge tracts of land being bought up by import-dependent countries such as China, India and Saudi Arabia, or the food-export bans that over thirty countries had imposed in an attempt to reduce prices at home and contain growing civil unrest.

As I presented my research on why food prices had been rising so much, I looked up at the audience and saw a familiar face listening intently. For some reason I felt uneasy, but I wasn't sure why. As I continued speaking, I kept trying to place the face I'd seen. Who *was* that guy? And why did seeing him there induce this weird queasiness in me?

When I saw him again during the coffee break, the penny dropped. The man in the audience was Nick Griffin, at that time the leader of the British National Party, Britain's fascist movement. Now I was properly spooked. Why had Nick Griffin come to listen to people like me at a conference on resource scarcity? Was the research that I and others like me were doing on the issue in some way *useful* to them?

Historian Timothy Snyder would say so. In his book *Black Earth* he explores how history teaches us that the fear of resource scarcity – the belief that there isn't enough food or land to meet everyone's needs, and hence that there's no choice but to grab it and if necessary fight for it – provides the perfect soil in which fascism can germinate.

Hitler's concept of *Lebensraum* ('room to live') had its roots in Germany's experience of being blockaded in World War One, which led to acute worries about its food supply. To prevent this from ever happening again, Hitler set out to seize vast areas of eastern Europe and transport its produce to Germany, thereby lighting the touch paper that would lead to World War Two.

In Rwanda, too, the genocidal killing of half a million

Tutsis in 1994 was rooted partly in how the tiny country's population had spiralled without any significant expansion in crop yields, which had instead dwindled as a result of drought, erosion, overgrazing and soil exhaustion. Many of the Hutu *génocidaires* later admitted that the killing was not just because of ethnic hatred, but also to seize land.

It's not hard to imagine how fears over the growing impact of climate change, coupled with increasing inequity in access to natural resources, could lead to more such catastrophes in the future. The risk, as Snyder puts it, is that 'a developed country able to project military power could, like Hitler's Germany, fall into ecological panic, and take drastic steps to protect its existing standard of living.'[1]

Perceptions of scarcity have the power to induce panic like almost nothing else. Think of what happens when there's a run on a bank, when the fear proliferates that depositors' demands for withdrawals won't be met. Or when a crowd starts to believe that there isn't enough space for everyone and morphs into a panicked stampede.

A variation on a similar theme is the ability of leaders like Donald Trump in the US or Nigel Farage in the UK to play on fears of being overrun by immigrants. During the US presidential campaign of 2016, Donald Trump played consistently on 'enemy within' themes – describing Mexicans as rapists and murderers,[2] and calling for an outright ban on the ability of Muslims to travel to the US.[3] During the UK's Brexit campaign, meanwhile, UK

Independence Party leader Nigel Farage unveiled a poster showing lines of refugees that bore a pronounced resemblance to 1930s Nazi propaganda footage.[4]

Often, of course, perceptions of scarcity have nothing to do with the reality. Take the case of global food security, for example. There's plenty of food available to feed the world's 7 billion people, and there still will be even when the Earth's population reaches 9 or 10 billion. Sure, we'll need to make some changes along the way – drastically reducing the astonishing amount of food we waste, for instance, and shifting to diets that are a lot less resource-intensive – but the fact remains that food crises happen not because there isn't enough to go around, but because of panic, incompetence, or sheer injustice. As the economist Amartya Sen puts it, 'Starvation is the characteristic of some people not *having* enough food to eat. It is not the characteristic of there not *being* enough food to eat.'[5]

So one myth that clearly *won't* help us to confront our current crisis of unsustainability is the myth of collapse: the idea that we're inevitably heading for a massive ecological crash, and that the best guide to our shared future can be found within the pages of post-apocalyptic fiction.

Unfortunately, the collapse myth is becoming all too prevalent. Take, for instance, the book *Ten Billion* by Stephen Emmott, who also happens to be head of computational research at Microsoft. After a detailed and intelligent discussion of climate change, resource scarcity and population growth, Emmott reaches the following conclusion:

> We urgently need to do – and I mean actually do – something radical to avert a global catastrophe. But I don't think we will. I think we're fucked.[6]

Or take philosopher John Gray, who opines in his book *Straw Dogs* that:

> Humans are like any other plague animal . . . [mankind] seems fated to wreck the balance of life on Earth – and thereby be the agent of its own destruction.[7]

Even leading ecologist James Lovelock – beloved of environmentalists as the creator of the Gaia hypothesis (the idea that the Earth can be thought of as a single mega-organism) – is at it:

> The bell has started tolling to mark our ending . . . only a handful of the teeming billions now alive will survive.[8]

And if shopping is the ritual that goes with the myth that we are what we buy, then collapsitarianism has its own ritual too. It's called 'prepping'. Preppers aren't always very specific about what they're prepping for, but they have an acronym to provide the general idea: TEOTWAWKI. The End Of The World As We Know It.

The website of the American Preppers Network offers advice on everything from selecting which slingshot or UHF radio is right for you, to strategies for 'bartering for survival in a post-collapse society'.[9] The site also helpfully includes advertisements offering membership of exclusive 'safe havens' for when things start to unravel. As you browse through the website's discussion threads, it's hard to escape the impression that many of its members seem to be actively looking forward to the impending cataclysm.

But while it's easy to poke fun at the preppers for now, it's also easy to imagine how they might come to be seen in a rather different light as climate impacts start to multiply, especially after major disasters, and they gleefully claim to have been right all along. It's already clear, after all, that we're in for an extremely turbulent – and at points downright scary – ride over the coming decades. And that's why the collapse myth is so dangerous: because of the risk that, as the reality of climate change starts to sink in, it becomes a self-fulfilling prophecy.

Of course, it might be that a couple of centuries from now, hindsight will show that the preppers' gloomy prognoses were right. But if that does prove to be the case, it may well be in large part the *result* of collapse myths: because enough people believed them, we concluded that there was nothing we could do in the face of climate change, and so did nothing.

Myths are powerful creatures, after all. They *create* our reality as much as they describe it: as the novelist Terry Pratchett once put it, 'People think that stories are shaped by people. In fact, it's the other way around.'[10] And if the myths we reach for in conditions of stress and crisis are ones about overshoot and collapse, and we all start to act accordingly – competing for resources rather than cooperating, fragmenting rather than coming together – then that will itself determine where we're headed.

But there's also an alternative possibility: that our current moment of crisis and transition proves to be the catalyst for powerful renewal and innovation in our core myths. There is plenty of historical precedent for this. China's warring states period, for example, was a period of huge upheaval – but it also gave rise to Taoism and Confucianism.[11] In the same way, the Jewish captivity in Babylon produced Jeremiah, Ezekiel and Isaiah, while the fall of the Western Roman Empire set the stage for the growth of Christianity in Europe.[12] Closer to our own times, successive waves of bubonic plague – which killed up to one third of the population in some parts of Europe – set the stage for the Renaissance.

So what might such a process look like in our times?

PART TWO

MYTHS FOR A NEW CENTURY

6

A larger us

IF THERE'S A GO-TO GUY on the size of the *us* that each of us belongs to, then it has to be American scholar and journalist Robert Wright, the author of *NonZero: History, Evolution and Human Cooperation*.

The title of *NonZero* refers to the distinction that game theorists make between 'zero-sum' and 'non-zero-sum' interactions. In a zero-sum game, the fortunes of the players are inversely related, so if one side is winning then the other side must be losing. Tennis, chess, war: all zero-sum games. In non-zero-sum games, on the other hand, everyone's interests are aligned, so that everyone wins or loses together. For instance, Wright observes in *NonZero*:

In 1970, when the three *Apollo 13* astronauts were trying to figure out how to get their stranded spaceship back to earth, they were playing an utterly non-zero sum game . . . the outcome would be either equally good for all of them or equally bad.[1]

Wright's central argument in *NonZero* is that while history is by no means predetermined, it does have a basic direction – one that tends towards more and more non-zero-sum cooperation and higher and higher levels of social complexity.

He's not disputing that history is littered with badly governed societies, brutal wars, horrific genocides and preventable disasters. He *is* arguing that non-zero-sum outcomes tend to win out over zero-sum ones in the long run, and that this is why human history so far is the story of how 'people become embedded in larger and richer webs of interdependence'.

Certainly ours is an age rich in interdependence and non-zero-sum outcomes. While the twenty-first century may seem violent, as conflicts from all over the world are beamed on to our televisions and phones, we actually live in extraordinarily peaceful times by historical standards. As Stephen Pinker puts it, 'violence has been in decline over long stretches of history, and today we are probably living in the most peaceful moment of our species' time on earth'.[2]

Similarly, we are living through a golden age for progress on poverty. Between 1988 and 2008, the proportion of humanity living in extreme poverty halved, with the poorest third of humanity seeing their incomes rise by between 40 per cent and 70 per cent, and those of the middle third rising by 80 per cent.[3] The number of children who die before the age of five has halved in the last fifteen years, meaning that every day 17,000 children live who a decade and a half ago would have died.[4]

There's another dynamic at play in history, too, one that runs parallel to the increase in non-zero-sum cooperation: the expansion over time of the size of the *we* with whom we empathize. From itinerant bands of hunter-gatherers to chiefdoms, from city states to kingdoms, and on to modern nation states and the staggeringly diverse communities of affinity and ethnicity in today's globalized world, our 'empathy horizon' keeps on expanding. As Wright put it to me when I met him:

> My own governing myth is an empirically based cosmological myth in the sense of sweeping historical narrative about how we got where we are today: there's been this drift toward interdependence, and that has tended to involve a kind of moral progress, an ability to take account of more and more different kinds of people who you're interdependent with.

This isn't to deny that there's still plenty of selfishness and bigotry around, or that humans are as capable as ever of zero-sum behaviours. But there's a direction of travel. You see it, for example, in Martin Luther King's 1963 'Letter from a Birmingham jail':

> I am cognizant of the interrelatedness of all communities and states. I cannot sit idly by in Atlanta and not be concerned about what happens in Birmingham. Injustice anywhere is a threat to justice everywhere. We are caught in an inescapable network of mutuality, tied in a single garment of destiny. What affects one directly, affects all indirectly.[5]

Nor is the expansion of the 'empathy horizon' limited to pioneers like Martin Luther King or Nelson Mandela. When the International Red Cross or Save the Children launches an emergency relief appeal after a natural disaster on the other side of the world, the outpouring of global generosity that results is something that wouldn't – *couldn't* – have happened five centuries ago. Sure, that's because we didn't have the technology back then. But that's part of Wright's argument: the emergence of new technology is one of the things that enable larger and larger scales of non-zero-sum cooperation.

So, in one sense, the story arc of human history is, above all, the grand narrative of how humans have kept becoming part of a larger *us*, over and over again. The evolutionary biologist and theologian Pierre Teilhard de Chardin saw this process, which he called 'planetization', as directly comparable to evolution – indeed, as an extension of it. Humanity started out in a phase he called 'emergence', then spread out across the world into different cultures and nations in a process of 'divergence'; now, it's coming back together through globalization and communications technology in a process of 'convergence'.

Crucially, Teilhard believed that this process of convergence and planetization, of becoming part of an ever larger *us*, doesn't eradicate what makes individuals or cultures unique. On the contrary:

> In any domain – whether it be the cells of a body, the members of a society or the elements of a spiritual synthesis – *union differentiates*. In every organised whole, the parts perfect themselves and fulfil themselves . . . The more other they become in conjunction, the more they find themselves as self.[6]

Now, almost two decades into the new millennium, we're at the very edge of completing Teilhard's process of planetization: a unique moment on our journey towards ever greater interdependence, complexity, and non-zero-sumness. But it's also a moment fraught with risk.

On one hand, we're poised right on the cusp of a

genuinely global *us* – with a global social media network, a global library of knowledge, a global economy, global governance institutions, a global sense of who we are. On the other hand, we're also on the verge of an unprecedented disaster in which we allow climate change – or other areas where our technological know-how risks surpassing our ability to use technology wisely, like nuclear proliferation – to run out of control. As Wright says in *NonZero*:

> At this moment of history, when social organization approaches the global level, and technologies of destruction reach commensurate scope, the arrow of history starts to quiver. Though getting to this threshold has long been so probable as to border on the inevitable, whether we successfully pass through it is another question altogether. And, while I'm basically optimistic, an extremely bleak outcome is obviously possible.

In other words, *this could all still go either way*. The collapsitarians' predictions of catastrophe – an outcome of extreme zero-sumness, as Wright would put it – might yet be vindicated. Equally, we could be about to tip decisively towards seeing ourselves as part of a 7 billion *us*. This is the extraordinary drama of the moment in history that we inhabit. And the single factor that will do most to decide how we fare, as we face this test, may ultimately be which stories – myths – we reach for to explain the transition we're facing.

7

A longer now

OUR FUTURE HAS BEEN SHRINKING for a long time now. It's a trend perfectly captured by the engineer and designer Danny Hillis, who observed in 1994 that:

> When I was a child, people used to talk about what would happen by the year 2000. Now, thirty years later, they still talk about what will happen by 2000. The future has been shrinking by one year per year for my entire life.[1]

Now that we're into the new millennium, the future is lucky to get a look-in at all. True, governments occasionally manage to agree grand targets for long-term action: 2015, for instance, saw governments declare a new set of 'sustainable development goals' on poverty, inequality and environmental sustainability, with a deadline of 2030. So far, though, the new goals are a long way from actually driving policy.

The timescales that really drive how business leaders and politicians act, by contrast, are *far* shorter. It's a point not lost on Larry Fink, the chief executive of BlackRock, the world's largest asset manager by far. In February 2016, Fink told the CEOs of companies in which he invests that 'Today's culture of quarterly earnings hysteria is totally contrary to the long-term approach we need.'[2]

(Lest anyone accuse him of singling out business leaders and letting policymakers off the hook, he also added that 'In Washington and other capitals, long-term is often defined as simply the next election cycle, an attitude that is eroding the economic foundations of our country.')

All this has a particular relevance, of course, for climate change: the grandfather of long-term issues, which is in turn one of the key reasons why so many psychologists who work on climate change are so downbeat about our chances of responding to it in time. (As Tony Leiserowitz of the Yale Project on Climate Change Communication observes, 'You almost couldn't design a problem that is a worse fit with our underlying psychology.'[3])

But the problem extends a long way beyond climate change alone. More broadly, our time horizons are being revved into even shorter spans by the dizzying pace of technological innovation. It's a trend perhaps best captured by the futurist Ray Kurzweil, who has done more than anyone to popularize the idea of a technological 'singularity', the notion that 'the pace of change in our human-created technology is accelerating and its powers are expanding at an exponential pace'.

Kurzweil argues that within a few decades from now:

... information-based technologies will encompass all human knowledge and proficiency, ultimately including the pattern-recognition powers, problem-solving skills, and emotional and moral intelligence of the human brain itself.[4]

He continues:

The Singularity will represent the merger of our biological thinking and existence with our technology... There will be no distinction, post-Singularity, between human and machine or between physical and virtual reality.

Whether or not you buy Kurzweil's particular focus on artificial intelligence, there are plenty of other innovations about to turn the world upside down that will make recent history look positively sedate by comparison. 3D printing, self-driving cars, blockchain currencies, nanotechnology, the internet of things, geoengineering, robotics, synthetic meat: it's not hard to see why many people agree with Kurzweil's prognosis that the future is moving faster and faster towards some kind of event horizon that it's impossible to see beyond.

Yet Kurzweil's worldview has also been widely criticized, including by technology guru Jaron Lanier, who argues that the problem with it is that 'humans aren't the heroes . . . the dominant story is machine-centric. It's technological determinism.'[5] What's missing in this story of being sucked towards some kind of technological cliff edge is the idea of people as the prime actors, or the sense that we have agency and choice over how the future unfolds, at what speed, and for whose benefit.

The futurist and environmentalist Stewart Brand, meanwhile, is evangelical about how technology can help us solve today's most pressing issues. Unlike Kurzweil, he puts people front and centre in no uncertain terms: 'We are as gods, and might as well act like it.'[6] It's an intriguing observation, most of all because of its studied ambiguity.

On one hand, it can seem like a statement of the utmost hubris: a monumentally dangerous ego inflation in which, like Icarus, we become drunk with our technological prowess before the inevitable crash to earth. On the other hand, it's equally possible to read his words as recognition not only of the awesome creative power of our free will and technological ingenuity, but also of the need for profound humility and wisdom as their necessary counterparts.

Brand himself tends firmly towards the second view, something that's clearest when you read what he has to say about our perceptions of time. Far from displaying giddy excitement at the pace of technological acceleration, Brand is emphatic that we need to slow down and take a far longer view than we've been accustomed to doing in recent decades.

Together with Danny Hillis, Brand chairs the board of the Long Now Foundation, an organization devoted to returning us to thinking in generational timescales. They aim to practise what they preach: one of the first things you notice on their website is that they prefix years with an extra 0, so that 2017 becomes 02017. (As they helpfully explain, 'the extra zero is to solve the deca-millennium bug which will come into effect in about 8,000 years'.[7])

Nor are they limiting their work to seminars and research papers. They're also engaged in a monumental construction project of their own, the Clock of the Long Now, which began with another thought of Hillis's, back in 01995:

> I cannot imagine the future, but I care about it. I know I am part of a story that starts long before I can remember and continues long beyond when anyone will remember me. I sense I am alive at a time of important change, and

I feel a responsibility to make sure the change comes out well. I plant my acorns knowing I will never live to harvest the oaks.

I think it is time for us to start a long-term project that gets people thinking beyond the mental barrier of an ever-shortening future. I would like to propose a large (think Stonehenge) mechanical clock, powered by seasonal temperature changes. It ticks once a year, bongs once a century, and the cuckoo comes out every millennium.

As with the need to prompt people to think in terms of a larger *us*, the role of prompting people to situate themselves in a longer *now* is one that used to be played by religious institutions – themselves no strangers to epic building projects that encouraged people to think in long timescales. It's no coincidence that so many of the world's lengthiest construction projects were undertaken in the name of religion: England's York Minster took more than two hundred and fifty years to build, while the Mayan Chichén Itzá complex and Angkor Wat in Cambodia were both over four hundred years in the making.

Other institutions of great longevity also exhibit a flair for long-term thinking. Brand tells a story about New College at Oxford University, which begins with an entomologist poking at the antique wooden beams of the college's dining hall and discovering to his dismay that they are riddled with beetles. The entomologist duly reported this to the college's fellows, who began to worry about where on earth they would find new beams of a similar size and calibre. As Brand recounts:

> One of the junior fellows stuck his neck out and suggested that there might be some worthy oaks on the college lands. These colleges are endowed with pieces of land scattered across the country which are run by a college forester. They called in the college forester, who of course had not been near the college itself for some years, and asked him if there were any oaks for possible use. He pulled his forelock and said, 'Well, sirs, we was wonderin' when you'd be askin'.'
>
> Upon further inquiry it was discovered that when the college was founded, a grove of oaks had been planted to replace the beams in the dining hall when they became beetly, because oak beams always become beetly in the end. This plan had been passed down from one forester to the next for over five hundred years, saying, 'You don't cut them oaks. Them's for the college hall.'[8]

Today, the Clock of the Long Now is under construction, inside a mountain in western Texas. Its builders have avoided scheduling a fixed completion date – naturally – but they do plan to open it to the public when it's finished.[9] In effect, they're using a vast construction project to help catalyse a new story: a myth for the twenty-first century.

8

A better good life

THE THIRD DEMAND presented by our current crisis of unsustainability is that we find myths that help us to reimagine what we consider to be the good life, moving towards a society that doesn't measure our standard of living by how much we consume. Nearly fifty years ago, Robert Kennedy observed in a speech on the 1968 presidential campaign trail that:

> The Gross National Product does not allow for the health of our children, the quality of their education or the joy of their play. It does not include the beauty of our poetry or the strength of our marriages, the intelligence of our public debate or the integrity of our public officials. It measures neither our wit nor our courage,

neither our wisdom nor our learning, neither our compassion nor our devotion to our country. It measures everything, in short, except that which makes life worthwhile.[1]

Since then, various policy wonks have suggested that rather than measuring GNP or GDP, we should simply measure happiness. The Asian kingdom of Bhutan, for instance, uses a metric it calls Gross National Happiness;[2] commissions of the great and the good have been formed to compare different ideas for 'beyond GDP' indicators, like the Happy Planet Index;[3] even the UN General Assembly has adopted a resolution solemnly declaring that happiness is a good thing and that governments should work towards it.[4]

But as Jules Evans, the author of *Philosophy for Life and Other Dangerous Situations*, notes, there's a question lurking in the background of these policy agendas: is happiness really something you can measure?[5] While some say yes, pointing to 'hedonic' measures which ask people at regular intervals how happy they feel, Evans enquires sceptically whether everything that matters can really be graded on a scale of 1 to 10. 'Would you really measure daily how much you love your kids?'

Similarly, he continues, while the celebrated US psychologist Martin Seligman has attempted to identify five ostensibly quantifiable elements of happiness – positive emotion, engagement, positive relationships, meaning and accomplishments – something more fundamental is absent: 'what's missing in Seligman's vision of the good life is goodness'.

So, despite the current vogue for replacing GNP with happiness, the shift still struggles to capture those aspects of life that can't be quantified. More importantly, it assumes that the most important determinant of a good life is personal enjoyment – as opposed to service to others, or a sense of purpose.

The need for that sense of purpose – and its importance at a time when it can seem as if something fundamental is out of kilter in the world – was well captured by the journalist Michael Ventura in his seminal essay 'The Age of Endarkenment'.[6] Ventura's essay sets out a powerful reflection on adolescence as part of a 1989 collection of writing on the phenomenon of the 'global teenager', published at a time when the children born during the early 1970s were starting to come of age, ushering in the global 'youth bulge' that continues to expand today.

Ventura's essay concludes with his recollection of a New Year's Eve party at which a thirteen-year-old boy, the son of his then partner, suddenly started to sob. Ventura tried to find out what was wrong, and at last the boy burst out with a pure distillation of adolescent rage and angst: 'Everything is so fucked, it's all so fucked, what's the point, it's all so fucking fucked.'

Ventura doesn't attempt to contradict him. Instead, he comes back with a statement of raw purpose, telling him that we are living through a dark age in which, 'for reasons we can't comprehend, everything's being turned inside out, everything's imploding and exploding at once, and we can't stop it'. But, he continues:

> . . . that doesn't have to rob us of purpose; in fact it's the opposite, it implies a great purpose. That what each of us must do is cleave to what we find most beautiful in the human heritage – and pass it on . . . And that to pass these precious fragments on is our mission, a dangerous mission – that if you were going to volunteer for crucial, hazardous work, work of great importance and risk, this might be the job you drew. And it isn't a bad job at all. Actually, it's the best job.

Ventura's insight that night – that the need for a larger sense of purpose is one of the defining traits of adolescence – tallies neatly with an observation made by the scientist Duane Elgin, who likes to ask people how grown up they think humanity is as a species:

> When you look at human behaviour around the world and then imagine our species as one individual, how old would that person be? A toddler? A teenager? A young adult? An elder?[7]

Most people understand the question intuitively, he continues, and wherever he's asked the question around the world, 'people have responded in the same way: at least two-thirds say that humanity is in its teenage years'. Which makes sense, Elgin continues, when you reflect that teenagers tend to be reckless, rebellious, concerned with appearance and status, focused on instant gratification, and drawn to cliques built around an 'us versus them' mentality.

Most of all, he argues, adolescence is a time when others (parents, schools, churches and so on) are generally running things – but growing up means increasingly taking charge of our own lives. In much the same way, he continues, most of us today can often feel as if someone else is in charge (big business, government, the media), but as we move into the early adulthood of our species we'll find that 'maturity requires taking more responsibility and recognising that we are in charge'.

Elgin also observes that in cultures across the world, adolescence is the time of life during which initiation customs are observed. Such rites not only mark the transition to adulthood, but also involve an element of trial by ordeal: a set of stressful situations in which new ways of relating to others are learned and established. Ventura, too, is deeply interested in the idea of initiation as a necessary part of the transition process that is adolescence. Writing about the often extreme behaviours of contemporary teenagers, Ventura observes that:

We tend to think of this extremism in the young as something new, peculiar to our times, caused by pop or TV or the collapse of values. The history of our race doesn't bear this out, however. Robert Bly and Michael Meade, among others, remind us that for tens of thousands of years tribal people everywhere have greeted the onset of puberty, especially in males, with elaborate and excruciating initiations – a practice that plainly wouldn't have been necessary unless their young were as extreme as ours . . .

Unlike modern societies, he continues, tribal people in the past met the extremism of their young with an equal but focused extremism from adults. Far from running from this moment in their children's lives, the adults of these tribal societies celebrated it with rituals that had been kept secret from the young until that point – rituals that 'focused upon the young all the light and darkness of their tribe's collective psyche, all its sense of mystery, all its questions, and all the stories told both to contain and answer those questions'.

But in our own society, Ventura continues, we've forgotten how to initiate our young, much as we've lost our old myths. True, he allows, it's a long time since we lived in tribal societies. But until World War Two (and setting aside the odd revolution when things would bubble over), the extremism of youth was simply repressed by wider society.

After the war, though, something shifted. Adolescent energies long suppressed suddenly burst into the open, from rock 'n' roll to punk, from Vietnam protests to Occupy and the Arab Spring. As Ventura puts it, 'adolescence has literally *become* the cultural air we breathe'.

At the time Ventura was writing, issues like climate change and extreme inequality were just beginning to impose themselves on the global consciousness. Now, here we are a decade and a half later, facing them in their full furious majesty as the defining trends of the century ahead. As Duane Elgin argues, it's exactly the kind of trial by ordeal that the process of initiation centres on: one that requires us to draw on our deepest internal resources, and that will leave us transformed if we manage to navigate our way through it successfully.

It's a trial that requires us, ultimately, to *grow up as a species* and begin to assume the responsibilities of adulthood. And in order to do that, we'll need stories like the one Ventura told his partner's son in 1989, which can help us to understand the process we're engaged in and how we can rise to it. Myths.

9

Redemption

WHEN BRENÉ BROWN HAD A NERVOUS BREAKDOWN and started seeing a therapist, as she recounted to a British journalist, 'the therapist said to me, "You have to embrace your vulnerability," and I was like, "Screw that."'[1] She is from Texas, after all. But as she explains in one of the most viewed TED talks of all time, she's come to realize that there is real power in willingness to show vulnerability:

> It was funny, I sent something out on Twitter and on Facebook that says, 'How would you define vulnerability? What makes you feel vulnerable?' And within an hour and a half, I had 150 responses . . . Having to ask my husband for help because I'm sick, and we're newly married; initiating sex with my husband; initiating

sex with my wife; being turned down; asking someone out; waiting for the doctor to call back; getting laid off; laying off people. This is the world we live in. We live in a vulnerable world. And one of the ways we deal with it is we numb vulnerability . . .

We are the most in-debt . . . obese . . . addicted and medicated adult cohort in US history. The problem is – and I learned this from the research – that you cannot selectively numb emotion. You can't say, here's the bad stuff. Here's vulnerability, here's grief, here's shame, here's fear, here's disappointment. I don't want to feel these. I'm going to have a couple of beers and a banana nut muffin.[2]

While the examples of vulnerability that Brown picked up from her Twitter and Facebook surveys come mainly from the realm of personal life, there's also a deeper, more collective set of issues at play. James Hillman (until his death the unofficial dean of Jungian psychoanalysis) recognized this when he suggested that:

The depression we're all trying to avoid could very well be a prolonged chronic reaction to what we've been doing to the world, a mourning and grieving for what we're doing to nature and to cities and to whole peoples – the destruction of a lot of our world. We may be depressed partly because this is the soul's reaction to the mourning and grieving that we're not consciously doing.[3]

Most of us have become adept at tuning out the steady beat of news items reporting on worsening climate impacts, for example. Droughts or typhoons on the other side of the world; more records broken for heatwaves, rainfall, loss of Arctic sea ice; even the events happening right in front of us in the developed world, like Hurricane Katrina in 2005 and Hurricane Sandy in 2012, or the catastrophic flooding that afflicts the UK with increasing regularity. In all these cases, we find ways not to think about what's happening around us.

More broadly, we've become world class at avoiding facing up to the effects of our actions on other people and other species. Biologists now agree that the Earth is in the midst of its sixth mass-extinction event. As science writer Elizabeth Kolbert observes, those of us alive today are in effect deciding the course of future evolution without any real awareness of what we are doing: 'No other creature has ever managed this, and it will, unfortunately, be our most enduring legacy.'[4]

Meanwhile, as I am constantly reminded in Ethiopia, it's the poorest people in the poorest countries – the countries with the lowest emissions, whose populations bear the least responsibility for having created the problem of climate change – who are most exposed to the risks of a damaged and unstable climate. Extreme weather events are at an all-time high, and the international humanitarian relief system is creaking under unprecedented strain, with evidence suggesting that climate change was an aggravating factor in creating the worst humanitarian emergency in the world: the war in Syria and the massive refugee flows it has set off.[5]

Against this backdrop, grief is a wholly appropriate – and necessary – response. Yet it's also one that we hardly ever allow ourselves to feel: we're scared, perhaps, of the vulnerability and pain that would come with it. So how would we give grief its proper place, rather than trying to avoid it?

In large part, through myth. Some of the most powerful myths in the Old Testament are about another historical moment of existential dislocation and loss: the destruction of Jerusalem in 587 BCE at the hands of Babylonian invaders, and the subsequent exile of the Jews to Babylon.

The theologian Walter Brueggemann has observed that during these times, the prophets of Israel played three crucial roles.[6] One was to describe reality as it was: to force people to face up to what was actually happening, and why. (The prophet Jeremiah was so good at doing this that his particular brand of outrage was subsequently awarded its own noun: 'jeremiad'.)

A second role was to help people to face and deal with the despair that comes with looking reality full in the face – a role played above all during Israel's exile by the Book of Lamentations, which is full of sentiments like this:

How lonely sits the city that was full of people! How like a widow has she become, she who was great among nations! She who was a princess among the provinces has become a slave. She weeps bitterly in the night, with tears on her cheeks; among all her lovers she has none to comfort her; all her friends have dealt treacherously with her, and they have become her enemies.[7]

Third and finally, prophets were charged with giving hope for the future amid the carnage of the present, something that the prophet Isaiah did in particular during the exile with visions such as this one:

And your ancient ruins shall be rebuilt; you shall raise up the foundations of many generations; you shall be called the repairer of the breach, the restorer of streets to dwell in.[8]

Brueggemann observes that all three of these 'prophetic tasks' are relevant to our own particular moment in the early twenty-first century. But he also notes that while today's campaigners are often superb at the first and third of them, they're largely *terrible* at the second.

Only very rarely do most activists allow themselves to feel (much less show) grief about the issues they work on. Yet when grief *is* allowed to be expressed, it can be both powerful and cathartic. During a UN climate summit in

2013, for example, the usual bureaucratic tenor of the talks was breached when the Philippines' lead negotiator suddenly broke down as he described the impact of Typhoon Haiyan, which struck his home as the summit was taking place, killing more than six thousand of his compatriots.[9]

Abruptly, there was a whole different dynamic in the room. The dry, zero-sum interactions typical of the UN climate process were disrupted as the real-world impact of climate change suddenly intruded and empathy forced its way in.

Grief, then, is another area where we need myths to help us change how we think – and, in this case, recognize and find a way to express the pain we feel about climate change. But there's another, even more complex emotion where myths also have a crucial role to play: guilt.

Unlike the kind of grief that arises from, say, bereavement, the grief that comes from climate change is – particularly for those of us living high-consumption lifestyles in the developed world – bound up with the fact that we are the ones creating climate change and using other people's share of the planet's 'environmental space'. This leads us straight into some very delicate psychological territory.

Activists sometimes seem to believe that if they can just make everyone feel guilty enough about climate change or resource consumption, a campaign breakthrough will follow. But as psychologists know, guilt is an extremely poor motivator. The reality is that, deep down, we *already* feel guilty, and it's a big part of why we don't want to think about the problem.

One way we try to avoid guilt is simply by repressing it and pretending it's not there. Unfortunately, as Jung famously observed, shoving unpleasant or painful emotions to the unconscious part of our psyche – the 'Shadow', as he called it – has a tendency to do more harm than good. In an observation that could almost have been made with climate change in mind, he wrote:

> Only an exceedingly naïve and unconscious person could imagine that he is in a position to avoid sin. Psychology can no longer afford childish illusions of this kind; it must ensure the truth and declare that unconsciousness is not only no excuse but is actually one of the most heinous sins. Human law may exempt it from punishment, but Nature avenges herself the more mercilessly, for it is nothing to her whether a man is conscious of his sin or not.[10]

Alternatively, we may avoid our own feelings of guilt by projecting them on to others, for instance through employing the 'enemy narratives' that we saw earlier: stories that explain why the crises we're facing are all someone else's fault (Exxon, the 1 per cent, the Koch Brothers, Saudi Arabia, evil multinationals or whoever) and that implicitly refute the possibility of any blame on our own account.

But the blunt reality is that those of us in developed countries – together with the new middle class in emerging economies – *are* guilty.

We are all using substantially more than our fair share of the atmosphere's capacity to absorb greenhouse gas emissions. The sustainable amount of CO_2 that each of us can emit through our electricity and fuel use, travel, food consumption and so on, is between 1 and 2 tonnes. Compare that with the emissions of the average person in Europe (6.8 tonnes), China (7.2 tonnes), or the United States (16.8 tonnes) – or, for that matter, with the quarter-tonne of CO_2 emitted by the average citizen of a country that meets the UN's definition of 'least developed'.[11]

The same applies to consumption more broadly, too. If you add up the productive land around the world needed to sustain the UK's current consumption levels – to grow its crops, graze its livestock, supply its timber, support its buildings and infrastructure, and provide forests to absorb its emissions – the tally comes to 5.45 hectares for each of us, compared with 1.14 hectares for the average citizen of a low-income country like Ethiopia.

With the Earth's actual 'biocapacity' working out at 1.78 hectares per person, a figure that's declining as the world's population grows, it's not hard to do the maths. People in the world's poorest countries should be entitled to consume significantly *more* per capita than they do today. But with those of us in the rich world already consuming far beyond our fair share, the risk is that higher consumption in poor countries will push us over environmental tipping points on climate change, biodiversity loss and other issues. The only way to match sustainability with fairness is for all of us to stay within our fair share – implying that each of us in the UK needs to consume about one third of the energy and resources we do today.[12]

It is not hard to work out what happens if we refuse to live within our fair shares. The impacts of unsustainability, as we've seen, are felt disproportionately by the poorest people: the ones who live in the flimsiest housing, who don't have savings to cope when disaster strikes, who grapple daily with inadequate health care and insecure livelihoods, and who are also – let's say it again – the ones who've done least to create climate change or global unsustainability in the first place. As injustices go, it's hard to find one more outrageous than this.

As I write these paragraphs, I am sitting at my desk in Addis Ababa, where I have lived for the past four years. Outside the city, the country is in the grip of the worst humanitarian crisis since the 1984 famine, with 20 million people – more than one fifth of the country's population – in need of emergency assistance.

The crisis is readily attributable to climate change: it's the result of a drought caused by an unusually severe El

Niño cycle. Many of the victims of the drought will become part of the statistics that already put the annual death toll from climate change at 400,000 a year (a figure set to rise to 700,000 a year by 2030).[13]

But the crisis is also attributable to *me*. I drive everywhere in Addis Ababa. I eat meat most days. I fly (a *lot*). I buy a ton of stuff I don't really need. And for all that this is a form of guilt that's collective and that obviously doesn't derive from *intending* to hurt anyone, that doesn't stop it from being real. The guilt is there. We all know it. The real question is: what are we going to do about it?

And this is where twenty-first-century myths also need to speak of redemption. This is a deep idea, one that goes a long way beyond merely being 'sorry', or even being forgiven by the people we've wronged. Most fundamentally, it's about the need to atone for what we've done: an idea that extends beyond repentance to encompass the need to make amends and repair the damage done.

As we'll see in later chapters, the idea of atonement is one that our oldest myths have had plenty to say about. And it's also an idea that prepares the ground for the final theme that twenty-first-century myths need to speak of: restoration.

10

Restoration

IN THE NORTH OF ETHIOPIA, the province of Tigray resembles an African Arizona with its vast, dusty vistas, mesa stone formations, and cacti dotting the landscape. It's a fragile place, and when things go wrong for Tigray's farms, the results are catastrophic. This place was Ground Zero for Ethiopia's 1984 famine, in which nearly half a million people died.

But now, something astonishing is starting to happen. The *Guardian*'s environment editor John Vidal went to see it in 2014:

> Fifteen years ago the villages around Abrha Weatsbha in northern Ethiopia were on the point of being abandoned. The hillsides were barren, the communities, plagued by

floods and droughts, needed constant food aid, and the soil was being washed away.

Today, Abrha Weatsbha in the Tigray region is unrecognisable and an environmental catastrophe has been averted following the planting of many millions of tree and bush seedlings. Wells that were dry have been recharged, the soil is in better shape, fruit trees grow in the valleys and the hillsides are green again.[1]

The World Resources Institute's Chris Reij observed to Vidal that 'the scale of restoration of degraded land in Tigray is possibly unmatched anywhere else in the world'. Now, the process is slowly being replicated across fully one-sixth of the country. And while Ethiopia as a whole is currently in the grip of a vicious drought, as noted in the last chapter, the parts of the country in which environmental restoration has taken root are proving far more resilient.

Nor is Ethiopia's story unique. Something similar is happening in China's Loess Plateau. In ancient times, the plateau was an area of huge fertility, before becoming a desert through deforestation, soil erosion, over-grazing and overpopulation. Now, it's undergoing a transformation similar to Tigray's, one that has unfolded over the course of just a few years.

In West Africa, too, plans to create a barrier of trees five thousand miles long and thirty miles deep – a 'Great Green Wall' – along the southern edge of the Sahara are now coming to fruition, with the promise of holding back and eventually even reclaiming the desert.[2]

Stories like these – stories of rebirth in which whole landscapes and ecosystems come back from the dead – have a power that the arid language of 'sustainability', which sounds as though its aspirations go just a fraction beyond mere harm reduction, will always lack. These are stories in which we go much further than merely halting the damage we're doing. Instead, they're about healing, repairing, resurrecting environments and returning them to their natural state.

More subtly, these images of a world made whole again speak of what happens when the process of atonement – *at-one-ment* – is completed. This can apply to people as well as ecosystems. Look at the revolutionary new idea of 'restorative justice', which New Zealand's pioneering justice ministry describes as 'a process for resolving crime that focuses on redressing the harm experienced by victims'.[3] Victim and perpetrator come face to face at a meeting during which the victim expresses how the crime has affected them. In many cases, the effect is transformative for both sides, with the offender wanting to know how they can make amends for what they have done: a process, in other words, of atonement. It's an approach that moves beyond our societies' frequent obsession with retributive justice – vengeance – towards an outcome that can heal both victim and perpetrator.

I think that tales of restoration are just about the most powerful and resonant kind there are. They speak directly to a profound yearning in all of us, an instinct that while the world may be broken, it can also be made right again, and that this may at some level be *what we are here to do*.

Nowhere do we need these kinds of stories more than in the context of climate change. So far, our ambitions only extend as far as halting the damage before it reaches truly catastrophic levels – as, for instance, with current targets to keep global average warming to 1.5 or 2 degrees Celsius.

But what if we went further, and aimed to return the atmosphere to the condition in which our ancestors found it?

What if our intention were to bring concentrations of carbon dioxide in the air back down to pre-industrial-revolution levels? What if we perceived our primary job for the next few centuries as being to nurse the Earth back to health?

The fact that this idea seems such an impossibly distant notion is *exactly* why we need stories of real power – myths – to help us find our way there.

These, then, are the things we need our twenty-first-century myths to do.

We need them to change how we see our place in the world, by prompting us to think in terms of a larger *us* than ever before, one that is capable of achieving non-zero-sum outcomes at the global scale. We need myths that help to shift how we see our place in time, situating ourselves in a longer *now* that stands at the intersection of a deep past and an equally deep future. We need these myths to challenge us to aspire to a different good life, one based less on material consumption and more on having a profound sense of belonging and, above all, purpose.

We need them to speak of redemption, helping us to recognize and face both our grief and our guilt, but also to move past both, above all by showing us how to atone for what we have done and start to make things right again. And we need them to give us hope, by telling us of a future in which the process of atonement has been successfully carried out, leading to an outcome in which the communities and ecosystems that we have damaged are restored.

It is asking much of any myth to deal with all these themes – and there is no single myth that will work for everyone. Our history is littered with lessons about the disastrous consequences that follow when any group of people becomes certain that they alone have discovered a universally applicable truth (and it bears noting that secular worldviews are every bit as capable of this kind of prejudice as religions).

Equally, though, the 'flatland' cultural relativism of recent decades – which insists that everything is as valid as everything else, and that ultimately nothing is really right or wrong – is no solution either. We need myths that embody values as they tell us about our place in the world and who we are; and given that our myths will all be different, we need a way of reconciling them with one another.

This has the potential to be a powerfully creative and enriching process, as former New York University president John Sexton has recognized: 'center-to-center contact between societies offers the promise that each may discover what is authentic and vital not only about the other but also about itself'.[4] For that to happen, he continues, an in-depth dialogue is needed in which the participants seek to learn from one another rather than refuting one another's claims.

In that spirit – of enquiring rather than proselytizing, and comparing rather than evangelizing – Part Three sets out the particular myth that works for me. It's a myth that I've found profoundly resonant in helping me to think about our current crisis and what it calls for from us, the crew on the bridge of the ship during the greatest storm our species has ever faced.

PART THREE

THE EVERLASTING COVENANT

11

The real Indiana Jones

MARGARET BARKER MAY NOT *look* like the real Indiana Jones. She wears large spectacles and pinches her lip as she listens intently when asked questions. She's a lay preacher at her local Methodist church. She talks with great affection about the ducks she used to own and (still more so) her grandchildren.

None of which should deceive you about how formidable she is. Fluent in Hebrew and ancient Greek, Barker is probably the leading expert on Jerusalem's lost First Temple: the one that housed the Ark of the Covenant, immortalized on celluloid in *Raiders of the Lost Ark*; the one that was destroyed by the Babylonians in 587 BCE.

Most of all, Barker's work is about the worldview of the people who built the First Temple, and how it differed from those who came after them. She argues that this worldview, which she calls 'temple theology', was systematically suppressed in a cultural revolution just before the temple was destroyed, and was excised from many of the texts that form the Bible as we have it today. But, she continues, it's now increasingly possible to reconstruct that worldview if you're willing to do the necessary theological detective work.

As I read her books, four things intrigued me. First, that Barker's description of temple theology had a lot to say both about the environment and humans' role in caring for creation, and about social justice. Second, that temple theology was emphatic that God was feminine as well as masculine. Third, that the God who emerges from this description seems a lot more interesting and benign than the psychopath who stomps through much of the Old Testament, smiting his enemies, sending plagues and pestilence, and slaying entire generations of first-born children. And fourth, that temple theology presented 'a theory of the origin of evil which does not involve Adam and Eve and original sin, and the massive burden of human guilt on which the Churches have fed for so long'.[1]

Start with the environmental aspect. If you ever had to go to Sunday school as a kid, then you already know that most important things in the Old Testament start with a covenant. God seems to go along making them every few pages in the opening books of the Bible: first with Noah, then with Abraham, then with Moses, and a bit later with David.

Barker's starting point[2] – along with other academics such as Robert Murray[3] – is to argue that the people of the First Temple believed in an older, greater covenant than any of these: one formed at the very beginning of time, when the cosmos was first created. The purpose of this covenant – the Everlasting Covenant or Cosmic Covenant – was to hold creation together, with the sea in its proper place, the climate in balance, and the stars in their orbits. (*Star Wars* fans will spot the similarity with the Force, which, as Obi-Wan Kenobi observes, 'surrounds us and penetrates us; it binds the galaxy together'.)

Moreover, the feminine aspect of God was central in creating this covenant. According to the Book of Genesis, the process of creation began when 'the Spirit of God moved upon the face of the waters'. What you don't get in today's translations, though, is the fact that in the Hebrew source the verb 'moved' takes the feminine form. God's feminine side wasn't just present at the creation: She was the one who instigated it.

This feminine aspect of God – Wisdom, as She was called – was represented in Eden too. In the account in the Book of Genesis there are two symbolic trees in Eden: the tree of life (which is explicitly referred to as feminine in the Book of Proverbs[4]) and the tree of knowledge. The tree of life was the tree that Adam and Eve *were* supposed to eat from, whose fruit was 'true knowledge of the divine creation'.[5] Those who ate of Wisdom's fruit – whether in

Eden, or in the First Temple – were transformed, becoming angelic and shining with glory in the image of God. As Barker writes, 'In temple theology, *resurrection was not a post-mortem experience*.'[6]

So this is a radically different story from the Eden story as we understand it today: it is much more mystical, and God comes across in a much less vengeful light.

In the Book of Genesis, God seems poised to pounce on Adam and Eve as soon as they make any move to grow up. But a very different story emerges from the Book of Enoch – a text lost outside Ethiopia (where, unlike anywhere else in the world, it forms part of the Bible) until the eighteenth century and translated into English for the first time only in 1906, and which is a key source for Barker's work on temple theology. Here, God *intends* Adam and Eve to shine. As Barker writes:

> The story in Genesis says, in effect, that Adam was condemned to remain blind, and never to achieve his godlike state. The God depicted in Genesis seems to be frightened, or jealous, of man in his godlike state. Mortality is the only way to keep him in check. The Enoch tradition experienced God differently, and had no fear of the angelic life of the wise ones.[7]

The covenant was, as Barker summarizes, 'a vision of the unity of all things, and how the visible material world relates to another dimension of existence that unites all things into one divinely ordained system'.[8] Humans had a crucial role in maintaining this system – 'tilling and keeping' the garden, as the Book of Genesis puts it – which entailed not only loving kindness (*ḥesed* in Hebrew) but also justice and seeing reality as it truly is (*mišpaṭ*).

The result of acting wisely in accordance with these principles was the state called *šalom*: peace. But this was peace in a far larger sense than just the absence of violence. Instead, it referred to a state of perfect balance and harmony; the state that existed in Eden.

I love so many things about the myth of the Everlasting Covenant. For one thing, it abundantly passes the test of pointing the way towards a larger *us*. The covenant is the centrepiece of a story in which everything is connected, everything is alive, and in which justice and care for creation go hand in hand.

It's also a myth that has a lot to say about a longer *now*. Time is a recurrent theme, and not only in the seven days of creation (which were commemorated on the sabbaths that occurred one day per week and one year in every seven, when no crops were grown so that the land could rest). More fundamentally, the Everlasting Covenant was rooted in mystical timelessness, with Day One of the creation process understood not as the first of seven days of creation, but rather as the cosmic unity that existed *before* creation began, and which *still* exists in the invisible parts that underlie and hold together the visible parts.

Finally, the myth of the Everlasting Covenant points the way towards a better good life. Adam and Eve are what they consume, but not in the sense of acquiring more and more material possessions. Instead, they eat and then become Wisdom, which in turn allows them to fulfil their crucial purpose of safeguarding the covenant: king and queen of creation not in the sense of it being theirs to rule, but, as in C. S. Lewis's Narnia, in the sense of it being theirs to protect.

But there's also a darker side to this myth, one that speaks to grief, guilt, and the need for redemption and restoration: because the covenant can also be broken. When that happens, the consequences – for Adam and Eve, for us and for the environment around us – are catastrophic.

12

Things fall apart

IN THE BOOK OF ENOCH, there are two interwoven stories about the origin of evil. While they differ on the details, both are about a rebellion in heaven, which begins when 'Powers' – angels – distort and corrupt the Earth through pride and self-will.

In one story, a powerful angel who knows the secrets of creation comes to Earth and teaches humankind some of those secrets – which, used without wisdom, lead to the corruption of the Earth. In the second story, two hundred angels see the beautiful daughters of men from heaven, lust for them, and come to Earth to rape them, in the process fathering 'half-breed children who were monsters and demons.'[1]

These myths are radically different from the Adam and

Eve story in their account of how evil first comes into the world. In Genesis, evil arises from Eve's disobedience. This not only forms the basis for the idea of original sin, but also for centuries of discrimination against women by the Church, from early Church fathers like John of Damascus ('Woman is the daughter of falsehood, a sentinel of hell, the enemy of peace') to the orgy of torture and killing that the Church embarked on from the fifteenth century as it persecuted women it suspected of being witches. To this very day, the Catholic Church refuses to countenance ordination of female priests.

As Barker observes, 'For fifteen centuries at least, women have been the victims of the Adam and Eve story, and have been blamed for being the victims.'[2] In the long-lost Book of Enoch, by contrast, Eve is not to blame for the origin of evil. Instead, it first arises in heaven itself, with angels who come to Earth and create havoc.

Does the idea of fallen angels have relevance in our times? The late Walter Wink, an American scholar and social justice campaigner, thought so. As an accomplished theologian, Wink regularly came across 'Powers' – angels – in religious texts. As a pioneer in non-violent political activism, he linked the idea in his mind to the 'Powers That Be' that he himself encountered, such as the apartheid regime in South Africa. And the more he reflected on the link, the more profound he found it.

Every organization or corporation, every city or nation,

every church or club has its own culture, a set of intangible characteristics that Wink called a 'spirituality', no less real or important for not being visible. And like the angelic Powers described in the Book of Enoch, Wink argued that 'the spirituality that we encounter in institutions is not always benign. It is just as likely to be pathological'.[3] He continues:

> . . . the Powers That Be are more than just the people who run things. They are the systems themselves, the institutions and structures that weave society into an intricate fabric of power and relationships. These Powers surround us on every side. They are necessary. They are useful. We could do nothing without them . . . But the Powers are also the source of unmitigated evils.
>
> A corporation routinely dumps known carcinogens into a river that is the source of drinking water for towns downstream. Another industry attempts to hook children into addiction to cigarettes despite evidence that a third of them will die prematurely from smoking-related illnesses. A dictator wages war against his own citizens in order to maintain his grasp on power . . .[4]

In other words, he continues, 'evil is not just personal but structural and spiritual'. It does not emerge only from the actions of individual humans, but also – and more fundamentally – from huge systems over which no individual has full control. Only by confronting the spirituality of an institution *and* its physical manifestations can the total structure be transformed.

For me, Wink's recognition that the most powerful forms of evil are not merely individual but *systemic* is crucial to understanding why myths about the covenant being broken have relevance to the real world.

Whenever wrongdoing has become truly pervasive – child abuse by Catholic priests, endemic fraud among investment banks in the run-up to the financial crisis, illegal phone tapping by British tabloid newspapers, legislators for sale in the US Senate or House of Representatives – there are always voices ready to claim that only a 'few bad apples' are guilty. But while it is true that there are guilty individuals in each of these cases, it is at the structural level that the evil is genuinely rooted. These are institutions whose 'angels' have fallen, leading them to become divorced from their purpose and cancerous in their impact.

The myths of the fallen angels in the Book of Enoch also tell of the consequences of the abuse of knowledge. In one myth, it is a fallen angel who instigates war among mankind: 'Azazel taught men to make swords, and knives, and shields, and breastplates.' In the other, the offspring of the fallen angels' rape of human women are 'great giants . . . and when men could no longer sustain them, the giants turned against them and devoured mankind'.[5]

Here, too, there is powerful relevance to our own times, in how we have become threatened by the 'great giants' of technological genius set loose from the protective bonds of wisdom, kindness and justice. It's no coincidence that when American physicist Robert Oppenheimer watched the first nuclear explosion in New Mexico in 1945, he reached instinctively for the language of myth, quoting Krishna's declaration in the Bhagavad Gita: 'I am become death, the destroyer of worlds.'[6]

Carl Jung, too, understood the risks of our technological capacities surging ahead of our capacity to use them wisely, writing in 1932 that:

> . . . the gigantic catastrophes that threaten us today are not elemental happenings of a physical or biological order, but psychic events. To a quite terrifying degree we are threatened by wars and revolutions which are nothing other than psychic epidemics.[7]

Not all of the knowledge that the fallen angels impart to humans – which in the Enoch myths includes writing, medicine and metallurgy – is necessarily bad, but it has the power to be used for immense evil when divorced from wisdom. As Barker observes, this is what is described in symbolic yet vivid terms in the rape of human women by the rebellious angels:

The urge to power shows itself in the abuse of wisdom, which is the feminine, creative aspect of God, and this manifests itself on earth in an exactly parallel form, the abuse of women and the corruption of the creation. In Hebrew the verb 'know' covered both knowledge and sexual activity.[8]

When the covenant is broken through ignorance or abuse of knowledge, the result is not liberation but disaster. The whole of creation is at risk of becoming distorted, or even of unravelling altogether. W. B. Yeats might have been thinking of such circumstances – or for that matter of today's 'culture wars', with societies fragmenting into strident worldviews that talk shrilly past each other – when in the aftermath of World War One he wrote:

> Things fall apart; the centre cannot hold;
> Mere anarchy is loosed upon the world,
> The blood-dimmed tide is loosed, and everywhere
> The ceremony of innocence is drowned;
> The best lack all conviction, while the worst
> Are full of passionate intensity.[9]

Nor are the consequences of the covenant being broken limited to humans, for the 'corruption of the creation' that results is also about environmental devastation. The Everlasting Covenant was what kept the Powers of the weather – forces like the winds, clouds, snow, frost, lightning, and the seasons themselves – in balance. Once the covenant is broken and the Powers slip their bonds, the results described in texts from two and a half millennia ago can start to seem eerily prescient in terms of the climate change and mass extinction taking place all around us today. One passage in the Book of Isaiah, for example, reads:

> There is no faithfulness, no love, no acknowledgement of God in the land . . . Because of this the land dries up, and all who live in it waste away; the beasts of the field, the birds in the sky and the fish in the sea are swept away.[10]

Meanwhile, the sea – separated from the Earth at the time of creation – is also suddenly free to transgress its boundaries, as if in anticipation of sea-level rise in the twenty-first century. Most vividly of all, the seasons slide out of kilter when the covenant is breached.

Although we've largely forgotten these warnings in their original scriptural settings, it's curious to notice how the underlying myths continue to bubble up in new contexts, something I'm reminded of every time I sit down to watch *The Lion, the Witch and the Wardrobe* or *Frozen* with my six-year-old daughter Isabel and three-year-old son Kit. In both of these modern-day covenant myths, the most obvious indication that something has gone terribly wrong is that the seasons aren't changing as they should; instead, the world is stuck in a winter that never ends.

In both cases, the cause of this is abuse of powerful knowledge that is being used for selfish ends rather than in the service of love – whether by the White Witch in Narnia, or by Elsa in *Frozen*. Which leads to the most important question of all, and the one at the heart of both of these stories and (as we'll see) many others besides: how do you mend a broken covenant?

13

And all shall be well

DURING THE FIRST TEMPLE period, the rituals
undertaken on the Day of Atonement, the most important
day of the religious year, were explicitly about healing
breaches in the covenant. These rituals seem obscure to us
more than 2,500 years later, but they were rich in
symbolism. They recognized the reality of collective guilt
and the threat this represented to creation, and the need
for atonement to resolve it. More than that, the rituals were
emphatic, in symbolic terms, about the necessity of *sacrifice*
as a means of dealing with collective sin and restoring
creation to its rightful state.

This is an insight that the most visionary campaigners
of our times have also recognized. Martin Luther King, for
example, said that 'Human progress is neither automatic

nor inevitable . . . Every step towards the goal of justice requires sacrifice.'[1] Gandhi, meanwhile, observed that 'the willing sacrifice of the innocent is the most powerful answer yet conceived by God or man to insolent tyranny'.[2]

It's also striking to notice how many of the most resonant and popular stories in contemporary fiction or cinema likewise reach their dramatic zenith when the protagonist overcomes evil not through might or victory, but instead through an act of heroic self-sacrifice in the name of love that then ushers restoration into the world. Consider:

- In *Star Wars*, Obi Wan raises his lightsabre and refuses to fight Darth Vader – an act by which, he warns, he will become 'more powerful than you can possibly imagine'. Luke Skywalker does the same at the end of *Return of the Jedi*; so does Han Solo at the denouement of *The Force Awakens*.

- In *The Lion, the Witch and the Wardrobe*, Aslan gives himself over to the White Witch to be killed before being resurrected as a result of what he calls a 'deeper magic' founded in 'the stillness and the darkness before Time dawned' (on Day One of the creation, in temple theology terms).

- In *Frozen*, Anna saves her sister Elsa by placing herself in the way of Prince Hans's sword blow – an 'act of true love' that redeems Elsa's power and places her magic knowledge back in the service of all, while also leading to Anna's own rebirth.

- In the final part of the Harry Potter series, Harry surrenders himself to Voldemort in the forest, ready to die in order to destroy the dark lord for ever, not yet understanding that Voldemort cannot kill him and that he will in effect be resurrected.

- In *The Lord of the Rings*, Frodo saves Middle Earth by grimly soldiering on towards Mount Doom at horrendous personal cost, while Gandalf sacrifices himself (and is later resurrected) to save the fellowship of the ring.

- And in Philip Pullman's *His Dark Materials* trilogy, Will and Lyra sacrifice their chance to be together in order to finish repairing the breaches between worlds – to repair, in other words, the fabric of creation and restore the covenant.

Many of these stories, of course, take their inspiration from the myth of Christ's death on the cross, most obviously those that involve resurrection – a debt that some of their authors (notably C. S. Lewis and J. K. Rowling) acknowledge explicitly.

The idea of resurrection through self-sacrifice is also prominent in the work of Carl Jung, who believed that everyone's central challenge in life was the task of individuation: making the psyche whole through assimilating the contents of the unconscious mind into conscious awareness.

For Jung, the path to achieving this task involved sacrificing the grasping ego and the fake personae that we assume in dealing with the world, in the process opening space for a whole, genuine expression of the true self. Just as atonement is how wholeness is restored to creation, individuation is how wholeness is restored to individual personality – with the idea of self-sacrifice central to both.

But does the idea of self-sacrifice have relevance today? This is where my enthusiasm for temple theology meets the political campaigner in me – for temple theology turns out to be *very* concerned with how we organize our economy and, above all, how it impacts on inequality, on poor people, and on the world around us.

The key idea here is that of *jubilee* – an ancient concept that has survived in the Bible we have today, and that can best be understood as the political and economic expression of atonement. Jubilees, and the closely linked idea of sabbaths, set out concrete procedures for how to correct economic, social and environmental imbalances – in effect, they were an instruction manual for how to build and maintain social and economic structures that protected rather than undermined the covenant.

Every seventh day, humans were supposed to rest and enjoy creation, as God had done when creation was finished. Sabbaths were about recognizing the sufficiency of creation, and all the abundance within it. This is the idea of a *different good life* made tangible – as Walter Brueggemann puts it, 'an act of both resistance and alternative . . . a visible insistence that our lives are not defined by the production and consumption of commodity goods'.[3]

Every seventh year, meanwhile, was a Sabbath year – a time of 'solemn rest for the land'.[4] No crops were sown; instead, people lived off whatever the land produced naturally. The land needed time to lie fallow so as to maintain its fertility. Environmental restoration was hence at the heart of sabbath years, and by extension jubilees too.

And finally, every seventh sabbath year – a year every half-century, in other words – was supposed to be a jubilee year. In addition to standard sabbath-year practices, land ownership would be reset, with houses and fields returned to their original owners.[5] (Given that the initial distribution of land between the twelve tribes of Israel was in proportion to their populations, the implication was that the jubilee reset of ownership would share land out anew on an equal per capita basis.)

Underlying this rule were a number of key principles, including that creation belonged to God, not humans; that two jubilees per century would prevent wealth inequality from building up over generations; and that 'every fifty years each family [would have] an opportunity to start afresh – free of debt and in possession of their own land'.[6]

Environmental restoration, cancellation of unsustainable debt, protection of the poor, prevention of the build-up of inequalities of wealth and opportunity across generations: an agenda that could have been written for the twenty-first century, and that in fact *has* been agreed by all of the world's governments in the form of the 2015 sustainable development goals, which I spent four years helping to negotiate at the United Nations and which are built around precisely these principles.

Throughout all of these rules for the correct observation of jubilee, the principle of sacrifice runs like a thread. Sacrifice of crop yields to allow the land to restore itself; sacrifice of work time to spend rest time with friends and family; sacrifice of debts owed, and of ownership of land.

So what happens as a consequence of these acts of atonement through sacrifice? In every case, whether in contemporary fiction or in the Bible, the result of these acts of self-sacrifice is restoration, with the covenant healed and creation brought back into balance. The Book of Enoch, for instance, describes a world in which the covenant has been repaired as follows:

> And then shall the whole earth be tilled in righteousness, and shall all be planted with trees and be full of blessing. And all desirable trees shall be planted on it, and they shall plant vines on it: and the vine which they plant thereon shall yield wine in abundance, and as for

all the seed which is sown thereon each measure of it shall bear a thousand, and each measure of olives shall yield ten presses of oil.[7]

Similarly, while the Bible begins with the fall from Eden and the loss of access to the tree of life, it ends in the final chapter of Revelation with the tree of life restored to Earth:

Then the angel showed me the river of the water of life, bright as crystal, flowing from the throne of God and of the Lamb through the middle of the street of the city; also, on either side of the river, the tree of life with its twelve kinds of fruit, yielding its fruit each month. The leaves of the tree were for the healing of nations.[8]

Now *that's* a myth.

PART FOUR

AND WE ALL LIVED HAPPILY EVER AFTER

A myth does not impart factual information, but is primarily a guide to behaviour. Its truth will *only* be revealed if it is put into practice – ritually or ethically. If it is perused as though it were a purely intellectual hypothesis, it becomes remote and incredible.[1]

Karen Armstrong

14

We are the battleground

SO HERE WE ARE – in a situation of extraordinary vulnerability. Our technological capacities have never been greater, yet we are struggling to come up with coherent responses to today's defining issues, whether climate change, inequality, migration or poverty. Our political discourse has fragmented into silo-like echo chambers with few common points of reference. Trust in mainstream leaders and institutions is waning.

I've argued that this situation has arisen in large part because of a myth gap in which our societies have lost their old stories without yet finding new ones. It's a situation that provides an ideal habitat for leaders like Donald Trump in the US, Marine Le Pen in France and Nigel Farage in the UK, for prophets of imminent social, environmental or economic collapse, for marketers who want to sell us the story that we are what we buy.

Too often, political progressives try to fight these hugely resonant stories with policy memos. Their hope appears to be (despite all the evidence to the contrary) that rational arguments and empirical data will win out against powerful narratives of 'us versus them', or corrupt politicians only out to line their own pockets, or conspiracies to falsify climate data. This was the mistake made by the ineffectual campaign to persuade Britons to remain in the EU. It was the mistake made by US climate campaigners in 2009, when they were routed by climate deniers and the Tea Party. And it was the mistake that paved the way for Donald Trump to win the US presidency in 2016.

If we want to stop making this mistake, and prevent these kinds of dark 'anti-myths' from filling the myth gap, then we need the *right* kind of myths – and fast. As I've said, what this means in practice is myths that can prompt us to think in terms of a larger *us*, a longer *now*, and a better good life; myths that can help us to process our sense of grief and guilt and take us towards redemption and restoration; and myths that can animate real-world social and political movements to incubate new values and push policymakers to make a radical shift towards justice and sustainability.

There is no single set of myths that will work for all of us. In a globalized society facing global challenges, we are far too diverse for that to be a viable option. Instead, it's for *each* of us to find the myths of regeneration and restoration that resonate for us personally – and then for *all* of us to find the areas of agreement between them, and sew together a quilt of compatible myths.

This will be much more challenging than simply agreeing to disagree about our core stories and values, as multiculturalism implies. Nor can it just mean a process of dialogue in which everyone has the opportunity to participate and be heard, but which at the end of the day struggles to reach any actual conclusions – a shortcoming that the Occupy Wall Street movement exhibited perfectly.

Instead, the challenge we face is to find genuine common ground and a hard-edged unity of purpose, while simultaneously recognizing and valuing what makes us different from each other.

So what does this look like in practice?

In the first instance, it means starting at the individual level and recognizing the wider impact of the choices we make in relation to the stories we use to make sense of the world around us. The most basic and fundamental responsibility for each of us is to make a conscious decision about which myths we adopt, rather than unconsciously allowing them to be chosen for us by the media, people around us, or leaders who play on themes of fear or anger.

This is, after all, one of the precepts that lie at the core of all the world's great wisdom traditions. It's clearly true of the mystics of every religion, who throughout the ages have emphasized the need for personal spiritual practice and stressed the need to take nothing on anyone else's authority. But it's also a principle that's easily found in non-theistic wisdom traditions: when the ancient Greek Stoics argued for the need for each of us to take responsibility for our thoughts as well as our actions, no reference was made to divine revelation.

Today, while formal religious observance may be in decline, interest in such forms of mental training is soaring. Take mindfulness meditation, which has been embraced by companies from Apple to Google and by institutions from the Pentagon to the British Parliament.[1] More than seven hundred mindfulness apps are now available for smartphones.[2]

The fact that interest in mindfulness has been growing steadily for decades suggests this is far more than just a fad. And significantly, the evidence base on mindfulness includes data showing that regular meditation has a pronounced positive effect in stimulating the anterior cingulate cortex – a part of the brain associated with the ability both to direct attention and activity in a purposeful way, and to manage impulsiveness and knee-jerk responses.[3]

In other words, mindfulness is a form of training that empowers practitioners to make decisions about how to respond to events rather than just acting reflexively, to practise discernment, and to see reality in a more considered, less reactive way. (From the perspective of the covenant myths discussed in Part Three, mindfulness can be seen as a way of practising *mišpaṭ* – of seeing reality as it truly is, and hence perceiving what has to be done to uphold the covenant.)

Nor is mindfulness the only option on offer. Stoicism is also alive and well in the twenty-first century, nearly two and a half millennia after its founding, with a vibrant international community of practitioners.[4] Cognitive behaviour therapy (CBT), too, is all about empowering people to make conscious decisions about how to respond to the world, and hence choose stories for themselves. But while the form that this kind of training can take is endlessly flexible, the underlying function – teaching ourselves to be aware of how we are interpreting the world around us – is unvarying.

The quantum physicist David Bohm recognized this when he observed, in an essay co-authored with two colleagues, Donald Factor and Peter Garrett, that:

> . . . with the aid of a little close attention, even that which we call rational thinking can be seen to consist largely of responses conditioned and biased by previous thought.

If we look carefully at what we generally take to be reality we begin to see that it includes a collection of concepts, memories and reflexes coloured by our personal needs, fears and desires, all of which are limited and distorted by the boundaries of language and the habits of our history, sex and culture. It is extremely difficult to disassemble this mixture or to ever be certain whether what we are perceiving – or what we may think about those perceptions – is at all accurate.

What makes this situation so serious, they continue, is that 'thought generally conceals these problems from our immediate awareness and succeeds in generating a sense that the way each of us interprets the world is the only sensible way in which it can be interpreted'. To overcome this, we need to slow things down, and develop the capacity to watch our thoughts critically rather than reflexively identifying with them.[5]

But if taking responsibility for the stories with which we choose to make sense of the world is the first step, then the second has to do with how we tell those stories to other people – and how we listen to what they say in return.

15

The social is political

As we saw in Chapter Three, politicians care obsessively about what ordinary people think, and most ordinary people make political decisions not on the basis of facts or data, but on the basis of what they hear from their friends, families, and other social networks.

This means that what we say in our ordinary lives has – in a small but very real way – potentially global implications. Conversations around the dinner table, at the office, in the pub and on social media are the ultimate determinants of the political space that politicians have to work within. They are creative in the rawest sense.

This is partly about the conversations we have on political issues, of course. But it's also, more fundamentally, about a rarer kind of conversation: the sort that unpacks and questions the underlying values, frameworks and stories that we use to make sense of political issues in the first place. Most of all, it's about having the courage to confront the status quo in these conversations, even if challenging existing social norms is never comfortable.

There's a particular power in being willing to talk about these issues by drawing on our own personal stories, and especially our doubts, fears, vulnerabilities and hurts. As we saw in Chapter Nine, most of us are reluctant to talk about personal life at all in work contexts, much less express difficult emotions like grief and guilt. But when we do, it becomes a different kind of conversation in which there's suddenly space to talk about the issues we usually ignore. The single trait that we most want to see in our leaders (and most often find lacking) is authenticity – and unless we're willing to bring the personal into conversation, we will almost always lack it ourselves.

But if speaking authentically is one half of the equation, the other half is how we listen. Today, the state of public debate in most developed societies is characterized by mutually exclusive tribes largely talking past each other. If we want to develop the capacity to find a sense of the common good through our diverse myths, this is what needs to change.

While training ourselves to be aware of our own thoughts and myths is crucial, then, so too is bringing a different attitude to how we engage in dialogue with others – which can also force us to re-examine the preconceptions inherent in our habitual myths. Bohm, Factor and Garrett describe the potential like this:

> [Dialogue] allows a display of thought and meaning that makes possible a kind of . . . immediate mirroring back of both the content of thought and the less apparent, dynamic structures that govern it. In dialogue this can be experienced both individually and collectively. Each listener is able to reflect back to each speaker, and to the rest of the group, a view of some of the assumptions and unspoken implications of what is being expressed along with that which is being avoided. It creates the opportunity for each participant to examine the preconceptions, prejudices and the characteristic patterns that lie behind his or her thoughts, opinions, beliefs and feelings, along with the roles he or she tends habitually to play. And it offers an opportunity to share these insights.[1]

This kind of dialogue tends to work best, they observe, within the context of relatively small groups – for very much the same reasons that small groups also provide the perfect launch pad for political movements, as we saw in Chapter Two. In groups of this size, as they explain, it is easier to work through anger, frustration and conflict when they inevitably arise – indeed, they can naturally become the central focus of the dialogue.

Participants find that they are involved in an ever-changing and developing pool of common meaning. A shared content of consciousness emerges which allows a level of creativity and insight that is not generally available to individuals or to groups that interact in more familiar ways.

What such dialogues can produce, in other words, is a synthesis in which personal myths are shared and examined, in the process building something new and larger – but without the participants losing the distinctiveness of the stories that make them individuals.

But there's a problem: in recent years, opportunities to have these kinds of conversation have been diminishing, in line with a wider trend of decline in social connectedness. Families eat together less often than they used to; the average American now eats one in every five meals in their car.[2] Outside the home, membership of groups of all kinds has been falling steadily, as the sociologist Robert Putnam has chronicled: from participation in parent-teacher associations, soccer leagues and reading groups to volunteering for Boy Scouts or the Red Cross, or being members of trade unions or political parties.[3] ' "We" became an "I", he concludes.[4]

With lower group participation and less social capital, there are fewer places to think collectively about our myths and values – a particular problem given that small groups have such a crucial role to play as the building blocks of long-lasting movements which have the power to incubate and diffuse new values. Where then can we look to kindle more of the deeper conversations we need?

Start with religions – the original myth curators and providers of spaces in which people can congregate and contemplate. While religious observance may be falling in the developed world, the decline is from a relatively high base. In the US, for instance, the percentage of religiously unaffiliated adults may have jumped from 16 to 23 per cent between 2007 and 2014 – but that still leaves 77 per cent who are religiously affiliated.[5] In the UK, meanwhile, 54 per cent of people identify themselves as religious.[6]

At one level, there are hopeful signs of a new focus on issues such as climate change and economic justice among religions, especially Christianity, the majority faith in most developed countries. Catholics – the largest branch of the largest religion – have become *far* more engaged with these issues, in particular following the publication of *Laudato Si: On Care for Our Common Home*, Pope Francis's 2015 encyclical on climate change and social justice.[7] Climate-change analysts credit this shift with having a galvanizing effect on leaders 'from almost every other faith', and with playing a critical role in changing the political context for the COP21 climate summit in December 2015.[8]

There are even some tentative signals of change in the habitual climate denial of many evangelical Protestants, especially in the United States. Recent polling data shows that an increasing number of rank-and-file evangelicals within the US don't see science and religion as being at odds with each other, while 'Creation Care' has emerged as a priority for some progressively minded evangelicals.[9]

Shifts like these are of course welcome if you're the kind of person who already cares about issues such as climate change or inequality – especially given that once religious groups are fired up, they're some of the most tenacious and effective campaigners out there. It's no coincidence that church congregations formed the backbone of so many past campaigns, from the nineteenth-century abolition of slavery by the British Parliament to the US black civil rights struggle of the 1950s and 1960s or the Jubilee 2000 Third World debt-relief campaign.

But at another level, what's *not* happening – yet – is the kind of dialogue that David Bohm wrote about. Pope Francis may be firing up Christians who were already progressively minded, but it's much less clear whether he's managing to win over more conservative Christians, above all where it really counts, in the US. Meanwhile climate writer David Roberts argues of the Creation Care movement's attempt to win over evangelicals:

> It failed. Pretty badly. The leaders of the movement were either drummed out of their leadership positions or

silenced by a furious backlash from other parts of the conservative coalition. Climate denial is now more firmly rooted among evangelicals than ever.[10]

So the need for meaningful dialogue remains urgent within Christianity and within other religions – and that's even before we get to the need for common ground *between* religions. As Hans Küng succinctly puts it in his writing about the need for what he calls a 'global ethic':

> No peace among the nations without peace among the religions. No peace among the religions without dialogue between the religions. No dialogue between the religions without investigation of the foundations of the religions.[11]

Most fundamentally of all, a furious clash between differing worldviews continues to simmer at the heart of public discourse. Religious believers disagree passionately with rational atheists – who turn out to be just as capable of fundamentalism as theists (as biologist Richard Dawkins regularly proves).[12]

Meanwhile, both religious believers *and* rationalists disagree passionately with relativists who dispute the idea of universal truth in the first place, instead arguing that all points of view are equally valid (apart, apparently, from those points of view that make claims to universalism, like religion and rationality), and stressing themes like pluralism, participation and multiculturalism.[13]

What remains absent, amid this clash of worldviews, is anyone willing to champion the argument that not only does each of these perspectives have something important to say at our current moment of crisis and opportunity, but also that a synthesis is possible, in which each worldview and set of myths is understood to be a vital part of the puzzle.

As George Marshall observed, climate change is – like the other defining issues of the early twenty-first century – 'far too large to be overcome without a near-total commitment across society'.[14] And we won't solve it if we're only prepared to talk and listen to people who start with the same values and worldview as us. So where might these spaces for meaningful dialogue open up?

16

The power of collective storytelling

WHILE RELIGION REMAINS a key reference point for the majority in most developed countries, the fastest growing cohort is the one that sociologists call 'nones' (people who reply 'none' when asked about their religion): a category that now includes one third of Americans under thirty, and 46 per cent of people in the United Kingdom.[1]

According to Robert Putnam, who refers to the same cohort as 'unknowns', the main reason for this shift is the combination of religion and politics:

> These were the kids who were coming of age in the America of the culture wars, in the America in which religion publicly became associated with a particular

brand of politics, and so I think the single most important reason for the rise of the unknowns is that combination of the younger people moving to the left on social issues and the most visible religious leaders moving to the right on that same issue.[2]

But while millennials may be turned off by organized religion, some argue that they *are* searching for other kinds of community in which they can explore questions of meaning and belonging – just the kinds of communities, in other words, that provide fertile soil for exploring and applying myths. Casper Ter Kuile and Angie Thurston, two friends who, notwithstanding being based at Harvard Divinity School, would firmly place themselves in the 'spiritual but not religious' camp, have explored this trend in their report *How We Gather*.[3] They argue that:

> ... when they say they are not looking for a faith community, millennials might mean they are not interested in belonging to an institution with a religious creed as the threshold. However, they are decidedly looking for spirituality and community in combination...
>
> The lack of deep community is keenly felt. Suicide is the third-leading cause of death among youth. Rates of isolation, loneliness, and depression continue to rise. As traditional religion struggles to attract young people, millennials are looking elsewhere with increasing urgency.

What's more, they continue, 'in some cases, they are creating what they don't find'. In their report, Ter Kuile and Thurston map out a range of new organizations that 'use secular language while mirroring many of the functions fulfilled by religious community'. In particular, they argue, these groups consistently emphasize some or all of six themes: community, personal transformation, social transformation, purpose finding, creativity and accountability.

Some of the organizations in which these traits can be found are national or even international in scale: Ter Kuile and Thurston note with keen interest the extraordinary levels of commitment displayed in fitness communities like CrossFit or SoulCycle, for example. Other communities are much smaller, and seek to create intimate gatherings where meaningful dialogue becomes possible, as for instance in the case of The Dinner Party, a network of twenty- and thirty-somethings who have all experienced a significant loss, who get together over homemade food to talk about it and how it affects their lives.[4]

It's significant, of course, that such small, high-commitment gatherings are very much the kind of 'little platoons' that have the power to add up to larger political movements, as we saw in Chapter Two. While they may not share the focus of religious groups on God or scripture, they *are* built around themes such as purpose, transformation and holding each other to account, and they *do* provide 'congregational spaces' in which deeper conversations are possible – which means they already have some of the key traits needed in order for dialogue to find common ground across individual myths.

These kinds of groups, then, provide one potential answer to the question of where conversations about myth and worldview could be explored. (It's also worth noting in passing that some of these member groups are starting to get into campaigning: Crossfit, for example, has engaged its groups in a sustained assault on 'big soda', demanding legislation that would mandate obesity and diabetes warnings on soft drinks, similar to the warnings on cigarette packets.)[5]

Another possible answer, meanwhile, can be found in cases where entire societies have engaged in collective storytelling exercises designed to imagine a range of possible futures and to think about the decisions that could lead to them.

In South Africa after the end of apartheid, for example, a broad mix of the country's political, business and civil society leaders came together for an intensive process of mapping out possibilities and choices that became known as the Mont Fleur Scenario Exercise.[6] While the participants came from extremely diverse backgrounds and had very different perspectives, they nonetheless succeeded in producing a shared map of where South Africa was and four different scenarios about how its future could unfold over the decade from 1992 to 2002. As Adam Kahane, who facilitated the conversations, summarized in 2007:

> 'Ostrich' told the story of a non-representative white government, sticking its head in the sand to try (ultimately in vain) to avoid a negotiated settlement with the black majority. 'Lame Duck' anticipated a prolonged transition under a weak government which, because it purports to respond to all, satisfies none.
>
> In 'Icarus', a constitutionally unconstrained black government comes to power on a wave of popular support and noble intentions and embarks on a huge and unsustainable public spending programme, which crashes the economy. In 'Flight of the Flamingos', the transition is successful, with everyone in the society rising slowly and together.
>
> These stories may not be relevant in 2007, but they reflected key choices facing South Africa in 1992, with particular emphasis on the nature of the political settlement and the economic policies that would follow.

As Kahane notes, what mattered most here was not the scenarios themselves so much as the quality and depth of conversation that led to them. While the Mont Fleur exercise involved relatively few people, it was typical of hundreds of informal forums that South Africans created during the transition to majority rule: some used scenarios, some didn't, but 'all created a safe·and open space'.

In the process, he continues, the forums developed a sense of *we* among the participants, together with an assumption of shared interest and identity and of being involved in a shared national project, at a time when 'the old was not yet dead and the new had not yet been born'. This in turn provided a crucial foundation for South Africa's 1994 political settlement and the transformation that followed.

More recently, the government of Singapore undertook a major collective storytelling process that involved over 47,000 Singaporeans in 660 dialogue sessions over the course of a year. The conversation focused explicitly on what kind of future Singaporeans wanted and how they should get there, and resulted in a clear synthesis of the perspectives that arose, and five core aspirations that Singaporeans wanted to pursue.[7]

Both of these cases underline how much innovation has taken place in collective storytelling in recent decades. And looking to the future, that pace of innovation is unlikely to slacken – particularly as we develop ways of applying new technology to the issue.

Already, social media has transformed the means by which we interact with each other. In one way, this has resulted in a highly tribalistic form of politics that has undermined the idea of a shared political discourse in society. But it is also at least *possible* that social media could provide a platform for the opposite dynamic, producing cohesion rather than fragmentation – especially when one starts to consider how technology will change the future of storytelling.

17

Technology and the future of myth

IF YOU HAD TO PICK one key moment for the emergence of modern myths about our place on Earth, it would surely have to be 1968, when William Anders, an astronaut on the *Apollo 8* mission, took an iconic picture of the Earth rising over the moon.[1] Environmentalists see that moment as a tipping point in our awareness of the planet as a single, fragile system, and the birth of the modern environmental movement.[2]

But if the effects of seeing the Earth from space were powerful for ordinary people who saw the photographs, they were far more dramatic for the astronauts who saw the real thing. The science writer Frank White has written at length about the 'overview effect' and how it has changed astronauts' perspective and sense of identity. Russell Schweickart, an astronaut on *Apollo 9*, put it like this:

You identify with Houston and then you identify with Los Angeles and Phoenix and New Orleans. And the next thing you recognize in yourself is that you're identifying with North Africa – you look forward to that, you anticipate it, and there it is.

And that whole process of what it is you identify with begins to shift. When you go around the Earth in an hour and a half, you begin to recognize that your identity is with that whole thing. That makes a change. You look down there and you can't imagine how many borders and boundaries you cross, again and again and again, and you don't even see them. There you are – hundreds of people in the Mideast killing each other over some imaginary line that you're not even aware of and that you can't see.

From where you see it, the thing is a whole, and it's so beautiful. You wish you could take one in each hand, one from each side of the various conflicts, and say, 'Look. Look at it from this perspective. Look at that. What's important?'[3]

Until now, only a tiny handful of astronauts have been able to experience this shift in perspective. But that's starting to change. Anyone who's watched 3D IMAX films of Earth from space, or of footage from the Hubble Space Telescope, will have had their breath taken away by the beauty of what they've seen. Anyone with a Twitter account can send an instant message to astronauts on the International Space

Station – and may even get a reply back. But we haven't seen anything yet – because this technology will within the next decade come to be seen as primitive in comparison with emerging virtual-reality technologies.

As Kevin Kelly, the founding editor of *Wired* magazine, observed in a long essay in May 2016, virtual reality (VR) – and especially *mixed* reality (MR), which overlays the virtual on to the real to create a composite image of the two – is now poised at the edge of a breakthrough in terms of the quality of its images and the experiences it can provide for the viewer.[4]

This breakthrough technology is about to allow ordinary people to experience the overview effect as intensely as astronauts do when confronted with the 'real thing' – or experience any number of other situations, from humanitarian disasters to melting ice caps, from total immersion in an ecosystem from the perspective of any given life form to seeing the whole process of evolution unfold in front of us.

If this sounds hyperbolic, then read Kelly's description of his own experiences of these new technologies. Visiting the suburban office of Magic Leap, a company at the cutting edge of VR and MR technologies, for instance, he describes how:

> . . . amid the low gray cubicles, clustered desks, and empty swivel chairs, an impossible 8-inch robot drone from an alien planet hovers chest-high in front of a row

of potted plants. It is steampunk-cute, minutely detailed. I can walk around it and examine it from any angle. I can squat to look at its ornate underside. Bending closer, I bring my face to within inches of it to inspect its tiny pipes and protruding armatures. I can see polishing swirls where the metallic surface was 'milled.' When I raise a hand, it approaches and extends a glowing appendage to touch my fingertip. I reach out and move it around. I step back across the room to view it from afar. All the while it hums and slowly rotates above a desk. It looks as real as the lamps and computer monitors around it. It's not . . .

As Kelly explains, what's being created here isn't simply a cutting-edge, much more immersive way of rendering computer-generated images. Instead, the nature of the internet itself will be rebooted by these new technologies. Today, he explains, the internet is a network of information. What VR and MR are building, by contrast, is an internet of *experiences*. He continues:

The recurring discovery I made in each virtual world I entered was that although every one of these environments was fake, the experiences I had in them were genuine. VR does two important things: One, it generates an intense and convincing sense of what is generally called presence. Virtual landscapes, virtual objects, and virtual characters seem to be there – a perception that is not so much a visual illusion as a gut feeling. That's magical. But the second thing it does is more important. The technology forces you to be present – in a way flatscreens do not – so that you gain authentic experiences, as authentic as in real life. People remember VR experiences not as a memory of something they saw but as something that happened to them.

The internet of experiences coming our way will transform just about every facet of life – but perhaps most of all, it will transform how we tell, and participate in, stories and myths. As Kelly describes, people working at the intersection of storytelling and technology – from Weta Workshop, who created the props for the *Lord of the Rings* movies and helped develop the culture of the Na'vi for *Avatar*, to science fiction author Neal Stephenson – are focusing their energies on VR and MR (both are working in collaboration with Magic Leap). For campaigners and myth-makers alike, a whole new realm is about to open up.

It's a point that was brought home to me vividly when I came across a description of how the United Nations started to use VR on the margins of the 2016 General Assembly, its annual gathering of heads of government. I've attended more than my fair share of these summits, and they always seemed to me to bring together the worst of the UN: its propensity to be a talking shop, where delegates talk past each other and posture for the media.

This year, though, something was different. Prompted by its first ever Creative Director, the UN was greeting national delegates as they arrived with VR goggles – which then transported them to meet refugees around the world. One of the UN team running the project observed that it was not uncommon to get the goggles back wet with tears.[5]

Most crucially of all, these new storytelling technologies will be available to all of us. The world wide web used to be somewhere where most of us read stuff while relatively few people actually created content; social media and the arrival of web 2.0 changed all that for ever. Film-making, too, used to be a linear process, with directors and producers making films and the rest of us watching them; now, anyone with a smartphone can be a film-maker. So as the new VR and MR technologies create whole new dimensions of storytelling and myth-making, they will also provide each of us with extraordinary powers to be the creators of those stories and myths.

As with any other technology, we will be free to use this power for good or ill. The importance of choosing wisely in the stories we tell each other will never have mattered more – especially at the moments when it appears as though everything is falling apart.

18

Eden 2.0

So what would happen if we used our powers of collective storytelling to imagine a future in which it all goes right, creating a myth about redemption and restoration that adds up, if you like, to an Eden 2.0. What would such a world look like?

To start with, each of us lives much more lightly than we do today. We buy less stuff. We throw less away. We eat less meat. We save and invest in ways aligned with the common good. We take more time over things.

On a larger scale, the world has completed the transition to a zero-carbon economy. Coal-fired power stations are museums now. All energy is clean. No one invests time and money any more in trying to find new sources of oil, gas and coal that we could never safely burn anyway.

With the damage halted, the repair work has started. Around the world, we are taking carbon out of the air and putting it back in the ground – in trees, in the soil, under the ground in reservoirs that used to contain natural gas. Greenhouse-gas levels in the air are falling and are heading back to where they were before the industrial revolution.

The number of species on Earth is rising steadily as extinct flora and fauna are reintroduced from DNA.[1] Bison once more roam the American plains. Agriculture is much more efficient and needs far less land, helped by more efficient diets.

Our economy is no longer at odds with environmental health. Instead, it's become the most powerful mechanism we have for restoring it. Tax systems have stopped taxing things we want (like work), and instead tax things we don't want (like pollution and resource use). As a result, companies compete on the basis of how small their environmental impact is.

As we live safely within environmental limits, we also share the Earth's common wealth fairly. Natural assets like land, water, minerals and the atmosphere are recognized as a shared inheritance belonging to everyone and no one. When a person or company uses that common wealth, they pair a fair price for doing so.

As well as keeping use of natural resources within sustainable limits, this also allows governments to redistribute the proceeds to the rest of society as a universal basic income paid to all citizens as a right (something Alaska was doing well before the turn of the millennium to distribute the proceeds of its oil wealth).[2]

With natural resources shared out equitably, they have ceased to be drivers of corruption, refugee flows and conflict. Instead, they are sources of shared prosperity. Nowhere is this shift more vivid than in the case of the climate, which is now protected by a safe global 'emissions budget' shared out on an equal per capita basis between all of the world's people. With poor countries able to sell their spare emissions quotas to higher emitters, they've secured a new revenue flow that has made aid flows an irrelevance.[3]

Partly as a result of this fairer approach to sharing natural wealth, absolute poverty no longer exists. No one is hungry. Everyone has access to clean water and sanitation, a quality education, adequate healthcare. Inequality is far lower than it was in the early twenty-first century. Everyone lives in peace.

Pipe dream? Perhaps. But also one that governments last year, for the first time, signed up to: for the future just described is essentially the one set out in the UN's 2030 Sustainable Development Goals (SDGs).

True, we're currently off track for every one of the SDGs, and bringing them within reach will require nothing short of a revolution in many cases. But as this book has argued, amazing things can happen when there's a mass movement with real purpose and a galvanizing story.

This Eden 2.0 is tantalizingly close. It could happen within our own lifetimes, and transform the lives of the next generations. But the path to it is unlikely to be direct.

For all that the collapsitarians risk creating self-fulfilling prophecies with their predictions of doom, they *do* have a point in their observation that the next few decades are likely to be defined by increasing shocks and stresses. Globalization has, after all, collapsed before, as my friend and colleague David Steven and I observed in a 2010 report we wrote for the think tank Chatham House:

> The early twentieth century, too, was defined by the ebb and flow of globalization. A hundred years ago, money, people, and ideas were able to pass freely across national borders – a state of affairs that appeared, as Keynes observed, 'normal, certain, and permanent, except in the direction of further improvement'. A few years later, when the First World War devastated Europe, it became clear how fragile the 'first globalization' had really been.
>
> As stresses like population growth, inequality, and global economic imbalances built up, so too did the price of failure. Technological change had dramatically increased the dangers of war, while European governments were locked into a political system that exacerbated rather than controlled tension. The stage was set for European governments to blunder into the Great War. With the subsequent Great Depression, and the era of protectionism, fascism, and communism that

followed, it took most of the rest of the century for globalization to recover.[4]

And that was *before* climate change. However fast we move to tackle global warming and shift to what might be called a restorative economy,[5] we're still in for an extremely turbulent few decades as a result of climate impacts that are now unavoidable. It's a period that's likely to be apocalyptic – in the precise sense of the word.

For 'apocalypse' doesn't mean 'the end of the world as we know it', although the collapsitarians would like us to think so. Rather, it means an unveiling of things as they really are, a *revelation*. And while the process is sometimes dramatic, sometimes frightening and sometimes painful, the result can be both cathartic and restorative.

The American writer Rebecca Solnit explores this phenomenon in her book *A Paradise Built in Hell*, which looks at the communities that form in the wake of natural disasters.[6] Whilst popular beliefs about what happens after disasters involve lootings, rapes, and the breakdown of law and order, she finds that the reality is much more often characterized by extraordinary selflessness, solidarity and sense of purpose, and that in these circumstances people are often astonished by how much more alive they feel than during the fragmentation and disconnectedness of ordinary life. She writes:

If I am not my brother's keeper, then we have been expelled from paradise, a paradise of unbroken solidarities. Thus does everyday life become a social disaster. Sometimes disaster intensifies this; sometimes it provides a remarkable reprieve from it, a view into another world for our other selves.

When all the ordinary divides and patterns are shattered, people step up – not all, but the great preponderance – to become their brothers' keepers. And that purposefulness and connectedness bring joy even amid death, chaos, fear, and loss.

Later in the same book, she continues:

A disaster sometimes wipes the slate clean like a jubilee, and it is those disasters that beget joy, while the ones that increase injustice and isolation beget bitterness – the 'corrosive community' of which disaster scholars speak. Some, perhaps all, do both. That is to say, a disaster is an end, a climax of ruin and death, but it is also a beginning, an opening, a chance to start over.

The stories and myths that we reach for in such moments are what determine whether we use those moments creatively or reactively, for a larger or a smaller *us*, for a longer or a shorter *now*, for a better or a worse idea of what constitutes a good life. As we relearn how to tell myths about where we are, how we got here, where we might be trying to go, and who we really are, we will discover extraordinary new capacities for creating the kind of future that we yearn for. As Isaiah once put it:

> . . . your ancient ruins shall be rebuilt; you shall raise up the foundations of many generations; you shall be called the repairer of the breach, the restorer of streets to dwell in.[7]

REFERENCES

Introduction

1 Bollier, D. *Power-curve society: the future of innovation, opportunity and social equity in the emerging networked economy.* Available online: http://www.aspeninstitute.org/policy-work/ communications-society/power-curve-society-future-innovation-opportunity-social-equity (accessed: 23/11/2014)

2 Bremmer, I. *Every Nation for Itself: What happens when no one leads the world.* New York: Portfolio (2013)

3 http://www.tearfund.org/~/media/Files/Main_Site/Campaigning/ OrdinaryHeroes/Restorative_Economy_Full_Report.pdf

PART ONE: THE FRONT LINE

1 How the climate movement rebooted itself

1 http://www.nytimes.com/1988/06/24/us/global-warming-has-begun-expert-tells-senate.html

2 1988: Kane, Parson poll for *Parents* magazine, USKANE.88PM7. RO98 and R11, data furnished by Roper Center for Public Opinion Research, Storrs, CT.

3 http://www.huffingtonpost.com/2013/04/09/margaret-thatcher-climate_n_3043873.html

4 Skocpol, T. 'Naming the Problem: What it will take to counter extremism and engage Americans in the fight against global warming' (2013). Available at: http://www.scholarsstrategynetwork.org/sites/default/files/skocpol_captrade_report_january_2013_0.pdf

5 http://grist.org/climate-energy/what-theda-skocpol-gets-right-about-the-cap-and-trade-fight/

6 http://www.globaldashboard.org/2013/08/12/on-course-for-3-6-5-3-degrees-celsius/

7 http://grist.org/climate-energy/what-theda-skocpol-gets-right-about-the-cap-and-trade-fight/

2   What climate activists learned from 2009

1 http://www.motherjones.com/environment/2014/10/climate-change-movement-peoples-march-wishful

2 An idea first coined by Tim Jackson and Ian Christie

3 Rochon, T. *Culture Moves*. Princeton: Princeton Press (1998)

4 https://labofii.wordpress.com/2015/08/23/drawing-a-line-in-the-sand-the-movement-victory-at-ende-gelande-opens-up-the-road-of-disobedience-for-paris/

5 http://w2.vatican.va/content/francesco/en/encyclicals/documents/papa-francesco_20150524_enciclica-laudato-si.html

6 http://p2pfoundation.net/Intrinsic_vs._Extrinsic_Motivation

7 http://www.nytimes.com/2015/11/06/science/exxon-mobil-under-investigation-in-new-york-over-climate-statements.html

http://www.washingtonpost.com/news/wonkblog/wp/2015/01/21/how-far-obamas-message-on-climate-change-has-come/

3 The problem with enemy narratives

1 http://www.climatechangenews.com/2015/05/08/george-marshall-we-need-to-engage-the-tories-on-climate-change/

2 https://www.micahmwhite.com/protest-is-broken/

3 Marshall, G. *Don't Even Think About It: Why Our Brains Are Wired to Ignore Climate Change.* London: Bloomsbury (2014)

4 http://www.winstonchurchill.org/resources/speeches/1940-the-finest-hour/their-finest-hour

5 *Ibid.*

4 The myth gap

1 Sachs, J. *Winning the Story Wars: Why those who tell and live the best stories will rule the future.* Cambridge, MA: Harvard Business Review Press (2012)

2 Armstrong, K. *The Battle for God.* London: Canongate (2000)

3 Jung, C. *Collected Works*, Vol. 5. New York: Pantheon Books (1953)

4 http://www.overshootday.org/about-earth-overshoot-day/

5 http://www.overshootday.org/

6 Heath, J. and Potter, A. *The Rebel Sell: How the counterculture became consumer culture.* Chichester: Capstone (2006)

7 http://storyofstuff.org/

5 Collapsitarianism

1 http://www.nytimes.com/2015/09/13/opinion/sunday/the-next-genocide.html?_r=0

2 http://www.vox.com/2015/7/8/8911467/donald-trump-immigrants-boycott

3 http://talkingpointsmemo.com/dc/donald-trump-muslim-ban-iowa-poll

4 http://www.theguardian.com/politics/2016/jun/16/nigel-farage-defends-ukip-breaking-point-poster-queue-of-migrants

5 Sen, A. *Development As Freedom*. Oxford: Oxford University Press (2001)

6 Emmott, S. *Ten Billion*. London: Penguin Books (2013)

7 Gray, J. *Straw Dogs: Thoughts on humans and other animals*. London: Granta (2003)

8 Lovelock, J. *The Revenge of Gaia: Why the Earth is fighting back and how we can still save humanity*. London: Penguin Books (2007)

9 http://www.americanpreppersnetwork.com/

10 Pratchett, T. *Witches Abroad*. London: Corgi (2013)

11 Armstrong, K. *The Great Transformation: The world in the time of Buddha, Socrates, Confucius and Jeremiah*. London: Atlantic Books (2007)

12 *Ibid.*; Stark, R. *The Triumph of Christianity*, New York: HarperCollins (2011)

PART TWO: MYTHS FOR A NEW CENTURY

6   A larger us

1 Wright, R. *NonZero: History, evolution and human cooperation: the logic of human destiny*. London: Abacus (2001)

2 http://www.slate.com/articles/news_and_politics/foreigners/2014/12/the_world_is_not_falling_apart_the_trend_lines_reveal_an_increasingly_peaceful.single.html

http://www.unl.edu/rhames/212/Stephen%20Pinker%20on%20the%20Decline%20of%20War.pdf

3 http://heymancenter.org/files/events/milanovic.pdf

4 http://www.un.org/millenniumgoals/childhealth.shtml

5 http://www.drmartinlutherking.net/martin-luther-king-jr-quotes

6 Teilhard de Chardin, P. *The Phenomenon of Man*. New York: HarperCollins (1959)

7 A longer now

1 Brand, S. *The Clock of the Long Now*: *Time and responsibility*. London: Basic Books (2000)

2 http://www.businessinsider.com/blackrock-ceo-larry-fink-letter-to-sp-500-ceos-2016-2

3 Marshall, G. *op. cit.* (2014)

4 Kurzweil, R. *The Singularity is Near*: *When humans transcend biology*. London: Duckworth (2005)

5 Lanier, J. *Who Owns the Future?* London: Allen Lane (2013)

6 Brand, S. *Whole Earth Discipline*. London: Penguin Books (2009)

7 http://longnow.org/

8 http://www.atlasobscura.com/places/oak-beams-new-college-oxford

9 http://longnow.org/clock/

8 A better good life

1 http://www.jfklibrary.org/Research/Research-Aids/Ready-Reference/RFK-Speeches/Remarks-of-Robert-F-Kennedy-at-the-University-of-Kansas-March-18-1968.aspx

2 http://www.grossnationalhappiness.com/

3 http://www.happyplanetindex.org/

4 http://www.nytimes.com/2012/03/29/opinion/the-un-happiness-project.html?_r=1

5 Evans, J. *Philosophy for Life and Other Dangerous Situations*. London: Rider Books (2012)

6 http://www.well.com/~mareev/TIMELINE/Sixties_great_writing/ Ventura-Age_of_Endarkenment.html

7 Elgin, D. *Promise Ahead: A vision of hope and action for Humanity's future*. New York: HarperCollins (2000)

9 Redemption

1 http://www.telegraph.co.uk/women/sex/self-help/9536381/Brene-Brown-on-the-power-of-vulnerability.html

2 https://www.ted.com/playlists/171/the_most_popular_talks_of_all

3 Hillman, J. and Ventura, M. *We've Had A Hundred Years of Psychotherapy – and the World's Getting Worse*. New York: HarperCollins (2009)

4 Kolbert, E. *The Sixth Extinction: An unnatural history*. London: Bloomsbury (2014)

5 https://www.americanprogress.org/issues/security/ report/2013/02/28/54579/the-arab-spring-and-climate-change/

6 Brueggemann, W. *Reality, Grief, Hope: Three urgent prophetic tasks*. New York: Wm. B. Eerdmans Publishing Company (2014)

7 Lamentations 1:1

8 Isaiah 58:12

9 http://america.aljazeera.com/articles/2013/11/11/philippine-representativeweepsatclimateconference.html

10 Jung, C. *op. cit.* (1953)

11 http://www.globalcarbonproject.org/carbonbudget/

12 WWF (2012) *Living planet report*. Available online: http://wwf.panda.org/about_our_earth/all_publications/living_planet_report/2012_lpr/ (accessed 10/11/2014)

13 http://www.thedailybeast.com/articles/2012/09/27/climate-change-kills-400-000-a-year-new-report-reveals.html

10   Restoration

1 http://www.theguardian.com/environment/2014/oct/30/regreening-program-to-restore-land-across-one-sixth-of-ethiopia

2 http://www.atlasobscura.com/articles/100-wonders-the-great-green-wall-of-africa

3 http://www.justice.govt.nz/policy/criminal-justice/restorative-justice

4 https://www.nyu.edu/about/leadership-university-administration/office-of-the-president-emeritus/communications/global-network-university-reflection.html

PART THREE: THE EVERLASTING COVENANT

11   The real Indiana Jones

1 Barker, M. *The Lost Prophet: The Book of Enoch and its influence on Christianity*. Sheffield: Phoenix Press (2005)

2 See e.g. Barker, M. *Creation: A Biblical Vision for the Environment*. London: Continuum (2010)

3 Murray, R. *The Cosmic Covenant*. London: Heythrop Monographs (1992)

4 Proverbs 3:18

5 Chartres, R. *Tree of Knowledge, Tree of Life*. London: Continuum (2004)

6 Barker, M. *op. cit.* (2005)

7 *Ibid.*

8 Barker, M. *op. cit.* (2010)

12   Things fall apart

1 *Ibid.*

2 Barker, M. *op. cit.* (2005)

3 Wink, W. *The Powers That Be: Theology for a New Millennium*. New York: Doubleday (1998)

4 *Ibid.*

5 Barker, M. *op. cit.* (2005)

6 Sachs, J. *op. cit.* (2012)

7 Jung. C. *Collected Works* 17, para 302

8 Barker, M. *op. cit.* (2005)

9 W. B. Yeats, 'The Second Coming'

10 Isaiah 24

13   And all shall be well

1 http://www.drmartinlutherking.net/martin-luther-king-jr-quotes

2 Gandhi, M. *Leadership* Vol. 10 No. 1

3 Brueggemann, W. *Sabbath as Resistance: Saying no to the culture of now*. Louisville: Westminster John Knox Press (2014)

4 Leviticus 25:4

5 Leviticus 25:23–24

6 Tan, K. *The Jubilee Gospel*. New York: Authentic Media (2008)

7 1 Enoch 10

8 Revelation 2 2:1–2

PART FOUR: AND WE ALL LIVED HAPPILY EVER AFTER

1 Armstrong, K. *op. cit.* (2009)

14 We are the battleground

1 http://uk.businessinsider.com/steve-jobs-zen-meditation-buddhism-2015-1?r=US&IR=T

http://www.fastcompany.com/3013333/unplug/3-reasons-everyone-at-google-is-meditating

http://parabola.org/2015/12/10/the-pentagon-meditation-club-by-tracy-cochran/

http://themindfulnessinitiative.org.uk/images/reports/Mindfulness-APPG-Report_Mindful-Nation-UK_Oct2015.pdf

2 http://lifehacker.com/the-best-mindfulness-apps-ranked-in-one-chart-1726392024

3 https://hbr.org/2015/01/mindfulness-can-literally-change-your-brain

4 http://www.philosophyforlife.org/category/stoicism/

5 http://www.david-bohm.net/dialogue/dialogue_proposal.html

15 The social is political

1 http://www.david-bohm.net/dialogue/dialogue_proposal.html

2 http://www.theatlantic.com/health/archive/2014/07/the-importance-of-eating-together/374256/

3 Putnam, R. *Bowling Alone: the collapse and revival of American community*. New York: Simon & Schuster (2000)

4 http://conversationswithbillkristol.org/transcript/robert-putnam-transcript/

5 http://www.pewforum.org/2015/11/03/u-s-public-becoming-less-religious/

6 http://www.lancaster.ac.uk/news/articles/2016/why-no-religion-is-the-new-religion/

7 http://w2.vatican.va/content/francesco/en/encyclicals/documents/papa-francesco_20150524_enciclica-laudato-si.html

8 http://www.ippr.org/juncture/high-pressure-for-low-emissions-how-civil-society-created-the-paris-climate-agreement

9 http://www.livescience.com/50162-most-evangelical-christians-dont-feel-hostile-to-science.html

https://www.lausanne.org/networks/issues/creation-care

http://www.tearfund.org/~/media/Files/Main_Site/Campaigning/OrdinaryHeroes/Restorative_Economy_Full_Report.pdf

10 See http://www.vox.com/2015/11/13/9729884/creation-care-evangelicals-climate-change

11 Küng Hans, *Islam, Past Present & Future.* Oxford: Oneworld Publications (2007)

12 http://www.scotsman.com/news/uk/an-embarrassing-fundamentalist-peter-higgs-scathing-verdict-on-richard-dawkins-1-2709488

13 For a much more detailed discussion of the nuances of these different worldviews, see e.g. Laloux, F. *Reinventing Organisations: A guide to creating organizations inspired by the next stage of human consciousness.* Brussels: Nelson Parker (2014)

14 Marshall, G. *op. cit.* (2014)

16 The power of collective storytelling

1 http://www.npr.org/sections/thetwo-way/2013/01/14/169164840/losing-our-religion-the-growth-of-the-nones

http://www.lancaster.ac.uk/news/articles/2016/why-no-religion-is-the-new-religion/

2 http://www.npr.org/sections/thetwo-way/2013/01/14/169164840/losing-our-religion-the-growth-of-the-nones

3 https://caspertk.files.wordpress.com/2015/04/how-we-gather.pdf

4 *Ibid.*

5 http://www.sandiegouniontribune.com/news/2015/nov/20/greg-glassman-crossfit-soda-label-crossfit-bill/

6 http://www.montfleur.co.za/about/scenarios.html

7 https://www.reach.gov.sg/read/our-sg-conversation

17   Technology and the future of myth

1 https://en.wikipedia.org/wiki/Earthrise

2 http://www.longfinance.net/images/reports/pdf/b1.pdf

3 White, F. *The Overview Effect*. Washington DC: American Institute of Aeronautics and Astronautics (1998)

4 http://www.wired.com/2016/04/magic-leap-vr/

5 http://thecreatorsproject.vice.com/blog/virtual-reality-unvr-refugees

18   Eden 2.0

1 http://reviverestore.org/

2 http://www.apfc.org/home/Content/aboutFund/aboutPermFund.cfm

3 http://www.cgdev.org/sites/default/files/SkyShares-PP-067.pdf

4 Evans, A. and Steven, D. *Organising for Influence*. London: Chatham House (2010)

5 http://www.tearfund.org/~/media/Files/Main_Site/Campaigning/OrdinaryHeroes/Restorative_Economy_Full_Report.pdf

6 Solnit, R. *A Paradise Built in Hell: The extraordinary communities that arise in disaster.* London: Penguin Books (2010)

7 Isaiah 58:12

ACKNOWLEDGEMENTS

This book was a long time in the making, and wouldn't have come about without the advice, encouragement and wisdom of a great many people.

Four mentors who have been indispensable both to the genesis of this project and to its unfolding have been Richard Chartres, Tom Spencer, Claire Foster-Gilbert and Margaret Barker – thank you all for your patient encouragement. My great thanks, too, to Paul Hilder, Graham Leicester, George Marshall, Kirsty McNeill, Debs O'Connor, Cynthia Scharf, Martin Stone, Casper Ter Kuile, and especially Emma Williams and Jules Evans for reassuring me at the right moments that this was a book worth writing.

I also probably wouldn't have written this without the

help of a wonderful group of people hosted by Avaaz for a week-long retreat in April 2015. I owe a great debt of gratitude to Robert Gass, Judith Ansara, Maura Bairley and Rachel Bagby for their ability to create such a special experience, and to Radhika Balakrishnan, Ian Bassin, Katherine Green, Dalia Hashad, Jamie Henn, Paul Hilder, Rajiv Joshi, Joanna Kerr, Adauto Modesto Jr, Luis Morago, Ricken Patel (to whom particular thanks for bringing us all together), Heather Reddick, Carolina Rossini, Esra'a Al Shafei, Theo Sowa, Tristram Stuart, and Farhana Yamin.

David Steven is one of the smartest people I know, has been a joy to write with (not to mention editing Global Dashboard together) over the last ten years, and has been a huge influence on my thinking on global governance and theories of influence, as well as keeping me sane during frequent bouts of frustration with the UN. I am also very grateful to have had the chance to co-author *The Restorative Economy* with Richard Gower, who is the source of many of the ideas in this book about movement building and values shifting, among other themes.

Thanks also to everyone at Tearfund – and especially to Sue Willsher, Paul Cook, Anna Ling, Matt Currey, Sam Barker, Katie Harrison, Madeleine Gordon and Jo Khinmaung-Moore, as well as Andy Atkins, Tom Baker, Richard Weaver and Laura Taylor – for their partnership over the last few years, for commissioning me to work on the *Restorative Economy* project, and for their consistently trailblazing work on climate change among British development NGOs.

My deep thanks to all the experts who agreed to be interviewed for this book, especially Bill McKibben, George Marshall, Jonah Sachs, Robert Wright, Jules Evans, Casper Ter Kuile and Angie Thurston. For advice on publishers I'm especially grateful to Tom Adams, Ele Fountain, Annabel Huxley and Deb O'Connor – thank you all for taking the time to wade through proposals and drafts. The Obi-Wan Kenobi quotation on p. 81 is from *Star Wars: Episode IV*, written by George Lucas.

It's been a great pleasure working with WhiteFox throughout the process, and a big thank-you to George Edgeller, John Bond and Tim Inman. I am especially grateful to Eleanor Rees for her immensely helpful edits to the earlier version of the book published as a Kindle Single on Amazon.

I owe huge thanks to Tim Smit for picking up the idea of publishing an updated version in this hardback edition, and for writing the foreword. Very many thanks also to Susanna Wadeson, Kate Samano and Patsy Irwin at Transworld and Mike Petty at Eden Project Books, and to my agent Richard Pike at Curtis Brown.

Finally, all the love and thanks in the world to my parents for always being ready with an encouraging word for some forty years now; to my brother and best friend Jules Evans for all the conversation, ideas and mischief over almost as long a period; and most of all to my wife Emma Williams, my partner in crime, conversation and long-distance train travel, as well as mother of the other two best things in my life, our kids Isabel and Kit. This book is for them.

Alex Evans is a Senior Fellow at New York University's Center on International Cooperation (CIC), with nearly twenty years' experience in climate and development policy. He joins Avaaz.org as a Campaign Director in 2017.

Alex has worked as Special Adviser to two UK Secretaries of State for International Development, as an expert on climate change in the UN Secretary-General's office, and with think tanks including Chatham House and the Brookings Institution. He has also worked as a consultant on futures and foreign policy for organizations from Oxfam to the US National Intelligence Council.

Alex lives in North Yorkshire and is married with two children.

Sir Tim Smit, KBE is Executive Vice-Chairman and co-founder of the award-winning Eden Project in Cornwall. His awards include the Royal Society of Arts Albert Medal (2003); 'Great Briton of 2007' in the Environment category of the Morgan Stanley Great Britons Awards; and a special award at the Ernst & Young Entrepreneur of the Year Awards (2011), which recognizes the contribution of people who inspire others with their vision, leadership and achievement.